Lecture Notes in Computer Science   14312

Founding Editors

Gerhard Goos
Juris Hartmanis

The series Lecture Notes in Computer Science (LNCS), including its subseries Lecture Notes in Artificial Intelligence (LNAI) and Lecture Notes in Bioinformatics (LNBI), has established itself as a medium for the publication of new developments in computer science and information technology research, teaching, and education.

LNCS enjoys close cooperation with the computer science R & D community, the series counts many renowned academics among its volume editors and paper authors, and collaborates with prestigious societies. Its mission is to serve this international community by providing an invaluable service, mainly focused on the publication of conference and workshop proceedings and postproceedings. LNCS commenced publication in 1973.

Ahmed Abdulkadir · Deepti R. Bathula ·
Nicha C. Dvornek · Sindhuja T. Govindarajan ·
Mohamad Habes · Vinod Kumar ·
Esten Leonardsen · Thomas Wolfers ·
Yiming Xiao
Editors

Machine Learning in Clinical Neuroimaging

6th International Workshop, MLCN 2023
Held in Conjunction with MICCAI 2023
Vancouver, BC, Canada, October 8, 2023
Proceedings

Springer

Editors

Ahmed Abdulkadir 🆔
Zurich University of Applied Sciences
Winterthur, Switzerland

Nicha C. Dvornek 🆔
Yale University
New Haven, CT, USA

Mohamad Habes 🆔
University of Texas Health Science Center at
San Antonio
San Antonio, TX, USA

Esten Leonardsen 🆔
University of Oslo
Oslo, Norway

Yiming Xiao 🆔
Concordia University
Montréal, QC, Canada

Deepti R. Bathula 🆔
Indian Institute of Technology Ropar
Rupnagar, India

Sindhuja T. Govindarajan 🆔
University of Pennsylvania
Philadelphia, PA, USA

Vinod Kumar 🆔
Max Planck Institute for Biological
Cybernetics
Tübingen, Germany

Thomas Wolfers 🆔
University of Tübingen
Tübingen, Germany

ISSN 0302-9743　　　　　　　　ISSN 1611-3349 (electronic)
Lecture Notes in Computer Science
ISBN 978-3-031-44857-7　　　　ISBN 978-3-031-44858-4 (eBook)
https://doi.org/10.1007/978-3-031-44858-4

This Springer imprint is published by the registered company Springer Nature Switzerland AG
The registered company address is: Gewerbestrasse 11, 6330 Cham, Switzerland

Paper in this product is recyclable.

Preface

The rise of neuroimaging data, bolstered by the rapid advancements in computational resources and algorithms, is poised to drive significant breakthroughs in clinical neuroscience. Notably, deep learning is gaining relevance in this domain. Yet, there's an imbalance: while computational methods grow in complexity, the breadth and diversity of standard evaluation datasets lag behind. This mismatch could result in findings that don't generalize to a wider population or are skewed towards dominant groups. To address this, it's imperative to foster inter-domain collaborations that move state-of-the art methods quickly into clinical research. Bridging the divide between various specialties can pave the way for methodological innovations to smoothly transition into clinical research and ultimately, real-world applications. Our workshop aimed to facilitate this by creating a forum for dialogue among engineers, clinicians, and neuroimaging specialists.

The 6th International Workshop on Machine Learning in Clinical Neuroimaging (MLCN 2023) was held on October 8th, 2023, as a satellite event of the 26th International Conference on Medical Imaging Computing & Computer-Assisted Intervention (MICCAI 2023) in Vancouver to continue the yearly recurring dialog between experts in machine learning and clinical neuroimaging. The call for papers was made on May 2nd, 2023, and submissions were closed on July 4th, 2023. Each of the 27 submitted manuscripts was reviewed by three or more program committee members in a double-blinded review process.

The sixteen accepted papers showcase the integration of machine learning techniques with clinical neuroimaging data. Studied clinical conditions include Alzheimer's disease, autism spectrum disorder, stroke, and aging. There is a strong emphasis on deep learning approaches to analysis of structural and functional MRI, positron emission tomography, and computed tomography. Research also delves into multi-modal data synthesis and analysis. The conference encapsulated the blend of methodological innovation and clinical applicability in neuroimaging. The proceedings mirror the hallmarks in the sections "Machine learning" and "Clinical applications", although all papers carry clinical relevance and provide methodological novelty.

For the sixth time, this workshop was put together by a dedicated community of authors, program committee, steering committee, and workshop participants. We thank all creators and attendees for their valuable contributions that made the MLCN 2023 Workshop a success.

Ahmed Abdulkadir

Deepti R. Bathula

Nicha C. Dvornek

Sindhuja T. Govindarajan

Mohamad Habes

Vinod Kumar

Esten Leonardsen

Thomas Wolfers

Yiming Xiao

Organization

Steering Committee

Christos Davatzikos   University of Pennsylvania, USA
Seyed Mostafa Kia   University of Tilburg, The Netherlands
Andre Marquand   Donders Institute, The Netherlands
Jonas Richiardi   Lausanne University Hospital, Switzerland
Emma Robinson   King's College London, UK

Organizing Committee/Program Committee Chairs

Ahmed Abdulkadir   Zürich University of Applied Sciences, Switzerland
Deepti R. Bathula   Indian Institute of Technology Ropar, India
Nicha C. Dvornek   Yale University, USA
Sindhuja T. Govindarajan   University of Pennsylvania, USA
Mohamad Habes   University of Texas Health Science Center San Antonio, USA
Vinod Kumar   Max Planck Institute for Biological Cybernetics, Germany
Esten Leonardsen   University of Oslo, Norway
Thomas Wolfers   University Clinic Tübingen, Germany
Yiming Xiao   Concordia University, Canada

Program Committee

Anoop Benet Nirmala   University of Texas Health Science Center San Antonio, USA
Matías Bossa   Vrije Universiteit Brussel, Belgium
Owen T. Carmichael   Pennington Biomedical Research Center, USA
Tara Chand   University of Jena, Germany
Niharika S. D'Souza   IBM Research, USA
Nikhil J. Dhinagar   University of Southern California, USA
Peiyu Duan   Yale University, USA
Benoit Dufumier   École polytechnique fédérale de Lausanne, Switzerland

Martin Dyrba   German Center for Neurodegenerative Diseases,
                                         Germany
Christian Gerloff   RWTH Aachen University, Germany
Hao Guan   University of North Carolina at Chapel Hill, USA
Xueqi Guo   Yale University, USA
Sukrit Gupta   Hasso Plattner Institute, Germany
Nicolas Honnorat   University of Texas Health Science Center San
                                         Antonio, USA
Stefanos Ioannou   Imperial College London, UK
Vinod Kumar   University of Tübingen, Germany
Kassymzhomart Kunanbayev   Korea Advanced Institute of Science &
                                         Technology, South Korea
Francesco La Rosa   Icahn School of Medicine at Mount Sinai, USA
Sarah Lee   Amallis Consulting, UK
Hao Li   Vanderbilt University, USA
Tommy Löfstedt   Umeå University, Sweden
Naresh Nandakumar   Johns Hopkins University, USA
John A. Onofrey   Yale University, USA
Amirhossein Rasoulian   Concordia University, Canada
Apoorva Sikka   Indian Institute of Technology Ropar, India
Lawrence H. Staib   Yale University, USA
Didem Stark   Charité - Universitätsmedizin Berlin, Germany
Ruisheng Su   Erasmus University Medical Center,
                                         The Netherlands
Elina Thibeau-Sutre   University of Twente, The Netherlands
Matthias Wilms   University of Calgary, Canada
Yiming Xiao   Concordia University, Canada
Tianbo Xu   Imperial College London, UK
Hao-Chun Yang   University Clinic Tübingen, Germany
Xing Yao   Vanderbilt University, USA
Mariam Zabihi   University College London, UK
Yuan Zhou   Fudan University, China
Zhen Zhou   University of Pennsylvania, USA

Contents

Machine Learning

Image-to-Image Translation Between Tau Pathology and Neuronal
Metabolism PET in Alzheimer Disease with Multi-domain Contrastive
Learning 3
    Michael Tran Duong, Sandhitsu R. Das, Pulkit Khandelwal,
    Xueying Lyu, Long Xie, Paul A. Yushkevich,
    Alzheimer's Disease Neuroimaging Initiative, David A. Wolk,
    and Ilya M. Nasrallah

Multi-shell dMRI Estimation from Single-Shell Data via Deep Learning 14
    Reagan Dugan and Owen Carmichael

A Three-Player GAN for Super-Resolution in Magnetic Resonance Imaging ... 23
    Qi Wang, Lucas Mahler, Julius Steiglechner, Florian Birk,
    Klaus Scheffler, and Gabriele Lohmann

Cross-Attention for Improved Motion Correction in Brain PET 34
    Zhuotong Cai, Tianyi Zeng, Eléonore V. Lieffrig, Jiazhen Zhang,
    Fuyao Chen, Takuya Toyonaga, Chenyu You, Jingmin Xin,
    Nanning Zheng, Yihuan Lu, James S. Duncan, and John A. Onofrey

VesselShot: Few-shot Learning for Cerebral Blood Vessel Segmentation 46
    Mumu Aktar, Hassan Rivaz, Marta Kersten-Oertel, and Yiming Xiao

WaveSep: A Flexible Wavelet-Based Approach for Source Separation
in Susceptibility Imaging 56
    Zhenghan Fang, Hyeong-Geol Shin, Peter van Zijl, Xu Li,
    and Jeremias Sulam

Joint Estimation of Neural Events and Hemodynamic Response Functions
from Task fMRI via Convolutional Neural Networks 67
    Kai-Cheng Chuang, Sreekrishna Ramakrishnapillai, Krystal Kirby,
    Arend W. A. Van Gemmert, Lydia Bazzano, and Owen T. Carmichael

Learning Sequential Information in Task-Based fMRI for Synthetic Data
Augmentation 79
    Jiyao Wang, Nicha C. Dvornek, Lawrence H. Staib, and James S. Duncan

Clinical Applications

Causal Sensitivity Analysis for Hidden Confounding: Modeling
the Sex-Specific Role of Diet on the Aging Brain 91
    Elizabeth Haddad, Myrl G. Marmarelis, Talia M. Nir, Aram Galstyan,
    Greg Ver Steeg, and Neda Jahanshad

MixUp Brain-Cortical Augmentations in Self-supervised Learning 102
    Corentin Ambroise, Vincent Frouin, Benoit Dufumier,
    Edouard Duchesnay, and Antoine Grigis

Brain Age Prediction Based on Head Computed Tomography Segmentation ... 112
    Artur Paulo, Fabiano Filho, Tayran Olegário, Bruna Pinto,
    Rafael Loureiro, Guilherme Ribeiro, Camila Silva, Regiane Carvalho,
    Paulo Santos, Eduardo Reis, Giovanna Mendes, Joselisa de Paiva,
    Márcio Reis, and Letícia Rittner

Pretraining is All You Need: A Multi-Atlas Enhanced Transformer
Framework for Autism Spectrum Disorder Classification 123
    Lucas Mahler, Qi Wang, Julius Steiglechner, Florian Birk,
    Samuel Heczko, Klaus Scheffler, and Gabriele Lohmann

Copy Number Variation Informs fMRI-Based Prediction of Autism
Spectrum Disorder 133
    Nicha C. Dvornek, Catherine Sullivan, James S. Duncan,
    and Abha R. Gupta

Deep Attention Assisted Multi-resolution Networks for the Segmentation
of White Matter Hyperintensities in Postmortem MRI Scans 143
    Anoop Benet Nirmala, Tanweer Rashid, Elyas Fadaee,
    Nicolas Honnorat, Karl Li, Sokratis Charisis, Di Wang,
    Aishwarya Vemula, Jinqi Li, Peter Fox, Timothy E. Richardson,
    Jamie M. Walker, Kevin Bieniek, Sudha Seshadri, and Mohamad Habes

Stroke Outcome and Evolution Prediction from CT Brain Using
a Spatiotemporal Diffusion Autoencoder 153
    Adam Marcus, Paul Bentley, and Daniel Rueckert

Morphological Versus Functional Network Organization: A Comparison
Between Structural Covariance Networks and Probabilistic Functional
Modes 163
    Petra Lenzini, Tom Earnest, Sung Min Ha, Abdalla Bani,
    Aristeidis Sotiras, and Janine Bijsterbosch

Author Index 173

Machine Learning

Image-to-Image Translation Between Tau Pathology and Neuronal Metabolism PET in Alzheimer Disease with Multi-domain Contrastive Learning

Michael Tran Duong, Sandhitsu R. Das, Pulkit Khandelwal, Xueying Lyu, Long Xie,
Paul A. Yushkevich, Alzheimer's Disease Neuroimaging Initiative,
David A. Wolk[✉], and Ilya M. Nasrallah[✉]

Perelman School of Medicine, University of Pennsylvania, Philadelphia, PA 19104, USA
{david.wolk,ilya.nasrallah}@pennmedicine.upenn.edu

Abstract. Alzheimer Disease (AD) is delineated by the presence of amyloid (A) and tau (T) pathology, with variable levels of neurodegeneration (N). While the relationship between tau and neuronal hypometabolism with positron emission tomography (PET) has been studied by T/N regression models, there has been limited application of image-to-image translation to compare between AD biomarker domains. We optimize a contrastive learning (CL) generative adversarial network for translation between paired tau and ^{18}F-Fluorodeoxyglucose (^{18}F-FDG) images from 172 older adults in an AD study. By comparing output domain query patches with positive and negative patches from input and target domains in latent space, the multi-domain CL loss approximates a cross entropy between latent output and a mixture of both original data distributions. Aligned with theory, CL model performance varies empirically based on use of dual vs. single domain CL loss, diversity of training data and negative patches, and CL temperature. Translation mapping yields realistic and accurate T and N images while reconstruction error validates regional relationships of T/N decoupling observed in regression models. Critically, biomarkers of non-AD pathology correlate with real and translated N images more than with input T images, suggesting our model imparts knowledge of non-AD influences on T/N relationships. Collectively, we present a proof-of-principle multi-domain CL approach to translate between clinical tau and ^{18}F-FDG PET, study mismatch between T and cellular responses and reveal patterns in AD vs. non-AD heterogeneity.

Keywords: Image Translation · Contrastive Learning · PET · Alzheimer Disease

M.T. Duong, S.R. Das and P. Khandelwal—Equal contribution.

Supplementary Information The online version contains supplementary material available at https://doi.org/10.1007/978-3-031-44858-4_1.

1 Introduction

Alzheimer Disease (AD) is a disorder of amyloid (A) and tau (T) accumulation, of which T is more strongly associated with neurodegeneration (N) [1]. Building on comparisons of the T/N relationship by regression models [2, 3], we employ image-to-image translation for T/N mapping between ^{18}F-Flortaucipir (tau) and ^{18}F-Fluorodeoxyglucose (^{18}F-FDG) positron emission tomography (PET). Deep learning enables prognostication from a single PET domain [4, 5] and multi-domain synthesis between amyloid and ^{18}F-FDG images [6]. Yet, the mapping of tau and ^{18}F-FDG PET with multi-institutional data and contrastive learning (CL) has not been studied.

CL is a powerful self-supervised method comparing latent data distributions [7]. Based on noise contrastive estimation (NCE), the InfoNCE objective was introduced as a loss comparing a query with positive and negative examples [8]. It has been shown that NCE objectives serve as a cross entropy between query and latent distributions [9, 10] and as a "distance" between distributions [11]. CL was integrated with a generative adversarial network (GAN) in the Contrastive Unpaired Translation (CUT) model [12]. Because CUT displays excellent conversion of unpaired images from domains with overlapping yet unique features, it is a promising tool for paired PET cross-tracer image translation where anatomy is similar, but uptake patterns differ.

Here, we demonstrate the utility of CUT for paired multi-domain image translation between T and N PET images. We develop a theoretical framework for CUT, which optimizes the sum of multi-domain, multilayer patchwise NCE losses, and in fact approximates a cross entropy between latent representations of the output distribution and a mixture of comparator data domains. From these mathematical foundations, we verify predictions on single vs. dual domain comparisons, diversity of training set and negative samples, and regularization. We analyze the mapping between T and N, which accurately portrays anatomical and uptake features. The reconstruction error between observed and translated ^{18}F-FDG images corroborates prior regression models of T/N mismatch. Importantly, the T/N mapping learns and reveals associations between N and non-AD copathology not seen on input T data. Overall, our translation approach displays utility in mapping the "match" and "mismatch" of T and N in AD.

2 Theory

We begin with an NCE loss. In Fig. 1, an encoder E maps from input X to latent space ($f_x: X \rightarrow Z_X$) and target Y to latent space ($f_y: Y \rightarrow Z_Y$). Generator G maps from latent space to output $\hat{Y}(g: Z \rightarrow \hat{Y})$ [12]. The Boltzmann CL loss $\mathcal{L}_{\text{contr}}(f_x, g, \mathbf{z}_{\hat{y}}, \mathbf{z}_x^+, \mathbf{z}_x^-)$ compares output samples $\mathbf{z}_{\hat{y}} \in \hat{Y}$ to input samples $\mathbf{z}_x^+, \mathbf{z}_x^- \in X$ with temperature τ [8],

$$\mathcal{L}_{\text{contr}}(f_x, g, \mathbf{z}_{\hat{y}}, \mathbf{z}_x^+, \mathbf{z}_x^-) = \mathbb{E}_{\mathbf{z}_x \sim p(\mathbf{z}_x)}\left[-\log \frac{e^{(f_x \circ g)(\mathbf{z}_{\hat{y}}) \cdot (f_x \circ g)(\mathbf{z}_x^+)/\tau}}{e^{(f_x \circ g)(\mathbf{z}_{\hat{y}}) \cdot (f_x \circ g)(\mathbf{z}_x^+)/\tau} + \sum_{i=1}^{M} e^{(f_x \circ g)(\mathbf{z}_{\hat{y}}) \cdot (f_x \circ g)(\mathbf{z}_{x,i}^-)/\tau}} \right]$$

(1)

where $\mathbf{z}_{\hat{y}}, \mathbf{z}_x^+, \mathbf{z}_x^-$ denote query, positive and negative samples, respectively, on a latent hypersphere via ℓ_2 norm. Similarly, $\mathcal{L}_{\text{contr}}(f_y, g, \mathbf{z}_{\hat{y}}, \mathbf{z}_y^+, \mathbf{z}_y^-)$ compares output $\mathbf{z}_{\hat{y}} \in \hat{Y}$ to target $\mathbf{z}_y^+, \mathbf{z}_y^- \in Y$. In a conceptual leap [12], $\mathcal{L}_{\text{contr}}$ was expanded to multilayer, patchwise

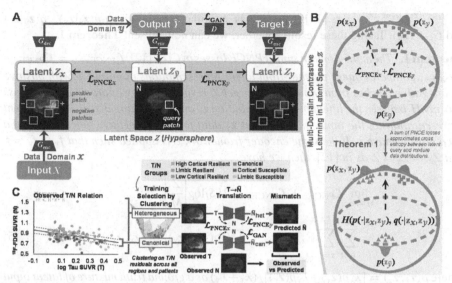

Fig. 1. Multi-domain PNCE objectives in CL translation of T and N maps, extended from [2, 3]. (A) CUT model architecture. (B) Hypersphere latent space with $\mathcal{L}_{\text{PNCE}x} + \mathcal{L}_{\text{PNCE}y} = \mathcal{L}_{\text{PNCE}x+y}$ approximating cross entropy H. (C) Outline of T→$\hat{N}$ (or N → $\hat{T}$) translation and mismatch.

NCE loss (PNCE) for L layers, S_l patch locations, and M negative samples, with

$$\mathcal{L}_{\text{PNCE}x} = \sum_{l=1}^{L} \sum_{s=1}^{S_l} \mathcal{L}_{\text{contr}}(f_x, g, \mathbf{z}_{\hat{y}}, \mathbf{z}_x^+, \mathbf{z}_x^-). \qquad (2)$$

An analogous $\mathcal{L}_{\text{PNCE}y}$ sums $\mathcal{L}_{\text{contr}}(f_y, g, \mathbf{z}_{\hat{y}}, \mathbf{z}_y^+, \mathbf{z}_y^-)$ to compare output and target. Since translation studies utilize GANs with a minimax loss from functions of generator G and discriminator D [12, 13], we have $\mathcal{L}_{\text{GAN}} = \mathbb{E}_{y \sim Y}[\log D(y)] + \mathbb{E}_{x \sim X}[1 - \log D(G(x))]$. CUT operates on the following total loss with parameters λ_x and λ_y set as 1, such that

$$\mathcal{L}_{\text{tot}} = \mathcal{L}_{\text{GAN}} + \lambda_x \mathcal{L}_{\text{PNCE}x} + \lambda_y \mathcal{L}_{\text{PNCE}y}. \qquad (3)$$

Herein, we provide theoretical claims with proofs adapted from literature and further developed in the **Appendix**. Extending results of [9, 10] from InfoNCE losses to PNCE losses, we find that optimizing a PNCE loss comparing query output patches with input/target patches in fact optimizes the cross entropy H (and hence KL divergence) between a latent output distribution $\mathbf{z}_{\hat{y}}$ and either of the latent representations of the data domains (input $\mathbf{z}_x$ or target $\mathbf{z}_y$) in a shared latent hypersphere in the limit of many negatives M. With latent distributions embedded on the hypersphere, we have

$$\lim_{M \to \infty} \mathcal{L}_{\text{PNCE}x} + LS_l \log\left|\frac{Z_x}{M}\right| = \sum_{l=1}^{L} \sum_{s=1}^{S_l} \mathbb{E}_{\mathbf{z}_x \sim p(\mathbf{z}_x)}[H(p(\cdot|\mathbf{z}_x), q(\cdot|\mathbf{z}_x))]. \qquad (4)$$

with ground-truth conditional distribution $p(\mathbf{z}_{\hat{y}}|\mathbf{z}_x)$, approximate conditional distribution $q(\mathbf{z}_{\hat{y}}|\mathbf{z}_x) = C_{f_x \circ g}(\mathbf{z}_x)^{-1} e^{(f_x \circ g)(\mathbf{z}_{\hat{y}}) \cdot (f_x \circ g)(\mathbf{z}_x)/\tau}$ and partition function $C_{f_x \circ g}(\mathbf{z}_x) = $

$\int e^{(f_x \circ g)(\mathbf{z}_{\hat{y}}) \cdot (f_x \circ g)(\mathbf{z}_x)/\tau} d\mathbf{z}_{\hat{y}}$ [10]. $\mathcal{L}_{PNCEy}$ has a similar relation given $p(\mathbf{z}_{\hat{y}}|\mathbf{z}_y)$, $q(\mathbf{z}_{\hat{y}}|\mathbf{z}_y)$ and $C_{f_y \circ g}(\mathbf{z}_y)$. Taking these results together, we can now present **Theorem 1**.

Theorem 1. *The multi-domain sum of PNCE losses in CUT* [12], $\mathcal{L}_{PNCEx+y} = \mathcal{L}_{PNCEx} + \mathcal{L}_{PNCEy}$, *compares query patches from latent representations of output $\hat{Y}$ with positive and negative patches from input X and target Y, respectively, by approximating a cross entropy H between the latent output distribution $\mathbf{z}_{\hat{y}}$ and the mixture distribution of latent representations of the input $\mathbf{z}_x$ and target $\mathbf{z}_y$ data for many negative patches M and a fixed non-negative τ. If the same encoder framework E is applied such that $f_x \approx f_y = f$ and $Z_x \approx Z_y = Z$ as in* [12], *then we have*

$$\lim_{M \to \infty} \mathcal{L}_{PNCEx} + \mathcal{L}_{PNCEy} + 2LS_l \log \left| \frac{Z}{M} \right|$$

$$= \sum_{l=1}^{L} \sum_{s=1}^{S_l} \mathbb{E}_{\mathbf{z}_x, \mathbf{z}_y \sim p(\mathbf{z}_x, \mathbf{z}_y)} \left[H \left(p(\cdot|\mathbf{z}_x, \mathbf{z}_y), q(\cdot|\mathbf{z}_x, \mathbf{z}_y) \right) \right]. \tag{5}$$

where $p(\mathbf{z}_x, \mathbf{z}_y) = [\lambda_x p(\mathbf{z}_x) + \lambda_y p(\mathbf{z}_y)]/(\lambda_x + \lambda_y)$ as a ground truth mixture of latent input and target distributions, $p(\cdot|\mathbf{z}_x, \mathbf{z}_y)$ is a ground-truth conditional distribution, $q(\cdot|\mathbf{z}_x, \mathbf{z}_y)$ is a conditional mixture approximation, and weights $\lambda_x = \lambda_y = 1$ for CUT [12].

Note **Theorem 1** is supported by findings in [10, 11] (**Appendix**). From our line of reasoning, we make several predictions about multi-domain translation with $\mathcal{L}_{PNCEx+y}$.

(i) A "dual" domain loss $\mathcal{L}_{PNCEx+y} = \mathcal{L}_{PNCEx} + \mathcal{L}_{PNCEy}$ optimizes 2 cross-domain comparisons and outperforms a single $\mathcal{L}_{PNCEx}$. While this was observed in [12], we offer more mathematical and empirical justification here. **Theorem 1** shows that 2 InfoNCE losses compare divergences between an output distribution with a mixture of both latent data distributions. In effect, this maximizes mutual information between output and both original data domains (Fig. 1). Incorporating features of both input and target domains, a dual-domain $\mathcal{L}_{PNCEx+y}$ favors more faithful data representations.

(ii) Changes to training data balance influence CL output quality [14]. Previous work has applied log-linear-based regression of tau and ^{18}F-FDG PET and clustering of residuals to identify patients with canonical (low residual) and non-canonical (large residual) levels of N relative to T [2, 3]. Here, we extend this result so specific models are trained on either canonical or heterogeneous cohorts. We predict the canonical trained models may be encouraged to mimic the typical relationship between tau and ^{18}F-FDG uptake, while the heterogeneous trained model may have better overall performance in capturing the variability in the tau PET and ^{18}F-FDG PET mapping.

(iii) More training variability in positive and negative sampling will improve performance. Heterogeneity of positive and negative patches can diversify feature learning and can be achieved by computing PNCE losses based on internal and external (int+ext) patches rather than internal alone (int). Indeed, internal and external training was shown to improve performance [15]. Since shortcut solutions suppress learnable features and impair instance discrimination [16], altering the source and diversity of negative samples may act as an implicit feature modification to raise discrimination.

(iv) Adjusting temperature parameter τ may impact regularization of performance in a setting of training data variation. We compare varying τ since changes in τ can affect CL performance [16]. Together, these predictions can be tested to translate between T and N PET domains and assess reconstruction error as T/N mismatch.

3 Methods

3.1 Patient Cohort and Imaging Data

We included participants from the Alzheimer's Disease Neuroimaging Initiative (ADNI) (http://adni.loni.usc.edu) with ^{18}F-Flortaucipir (tau) PET and ^{18}F-FDG PET performed <1 year of each other, along with a measure of amyloid status (amyloid PET or cerebrospinal fluid) and magnetic resonance imaging (MRI) (<1 year of PET scans). Of these, 172 amyloid-positive (A+) participants on the AD continuum were found: 102 had mild cognitive impairment (MCI), 62 had dementia and 8 were cognitively normal (**Supplementary Table 1**). Median time between ^{18}F-FDG vs. tau PET was 12 days (80% of cases within 1 month). Processed PET images with uniform isotropic resolution (8mm full-width-at-half-maximum) were obtained with the ADNI archive description "Coreg, Avg, Std Img and Vox Size, Uniform Resolution." ADNI MRI included a T1 structural scan (resolution $1.0 \times 1.0 \times 1.2$ mm^3).

3.2 Image Processing and PET Regional Analysis

MRI scans were processed using the ANTs pipeline for inhomogeneity correction, brain extraction, template registration, and cortical thickness measurement per published protocols [2, 3]. Brain MR images were divided into 104 Gy matter regions-of-interest with multi-atlas segmentation. PET images were co-registered to T1-weighted MRI with ANTs using rigid-body transformation. Standardized uptake value ratio (SUVR) maps were generated with reference regions specific to each radiotracer: inferior cerebellar cortex for ^{18}F-Flortaucipir and cerebellar cortex for ^{18}F-FDG [3]. Mean regional tau and ^{18}F-FDG measures were extracted from tau and ^{18}F-FDG SUVR maps. Methods for model regression and hierarchical clustering to form T/N mismatch groups are in the **Supplementary Methods** and Fig. 1C [2, 3].

3.3 Network Architecture and Implementation

The CUT model was trained and tested per [12]. Briefly, 3-dimensional SUVR maps were padded to $256 \times 256 \times 256$ voxels and sliced into 256×256 pixel slices along sagittal, axial and coronal planes. Images were also flipped randomly in training. These serve as data augmentation since multiview approaches improve generalization of instance discrimination [17–19]. Aligned data was preprocessed as crops and training batch size was 5. Training was run to 300 epochs and the number of epochs to decay was 100. Learning rate was 0.0002. Patches were 64×64 pixels, with 256 patches/image/iteration. GAN mode was least squares GAN and the architecture consisted of a 9-block ResNet generator and 3-layer PatchGAN discriminator. The encoder for PNCE loss was comprised of the generator encoder arm and a multi-layer perceptron with 2 layers [12]. CUT had more realistic PET output than pix2pix and CycleGAN.

Models were trained on 44 participants derived from the canonical group or the entire cohort to evaluate effects of training data diversity. As a more homogeneous dataset with "typical" relationships between T and N, a canonical training set had 44 randomly selected A+ participants categorized as canonical based on a previous clustering method [3] or were cognitively normal. Conversely, as a more diverse set representing variability within the T/N relationship, the heterogeneous training set had 44 randomly selected A+ canonical and non-canonical participants at roughly similar ratios in the cohort to ensure the set more faithfully captured the heterogeneity of the larger ADNI population and included patients from smaller mismatch groups.

Optimization experiments varied PNCE loss (single or dual losses), training set diversity (canonical or heterogeneous), negative sample diversity (internal+external or internal), and temperature τ. Models were tested on 50 randomly selected participants from all T/N groups unseen at training time and evaluated by the Fréchet Inception Distance (FID, lower is better), which can assess synthetic biomedical images [20, 21].

After optimization experiments, we generated translation output with 4-fold cross-validation on the whole set of 172 A+ participants with the best performing T→N̂ model settings. Outputs from 3 planes were combined by weighted average. For the canonical training set, each of the 4 folds consisted of 44 randomly selected A+ participants from the canonical group. For the heterogeneous training set, each of the 4 folds had 44 A+ participants selected via stratified random sampling to promote a diverse distribution of participants from each of the non-canonical groups (wherein smaller groups such as limbic resilient may not necessarily be represented in random sampling alone). Models were tested on remaining participants that were unseen during training of that fold, ensuring each participant was in the training set at least once and allowed calculation of average fold output SUVRs.

Non-AD copathology imaging measures were assessed by the cingulate island ratio (a marker of α-synuclein and Lewy body disease) and the inferior/medial temporal/frontal supraorbital ratio (I/MTL/FSO) (a marker of TDP-43 encephalopathy). Linear regressions were computed from non-AD markers and regional SUVRs [22, 23].

4   Experiments and Results

4.1   PNCE Theoretical Predictions Optimize the T/N Translation Mapping

First, we conducted a series of optimization experiments to assess the roles of training data imbalance and single comparison vs. dual comparison PNCE objectives on translation. (i) We posited the dual PNCE objective $\mathcal{L}_{PNCEx+y}$, which contrasts between output and a mixture of data distributions, outperforms single PNCE loss $\mathcal{L}_{PNCEx}$ contrasting between output and input distributions. In fact, models of $\mathcal{L}_{PNCEx+y}$ converged faster to a superior FID than models of $\mathcal{L}_{PNCEx}$ (Fig. 2A). (ii) We predicted training data diversity improves translation performance, while $\mathcal{L}_{PNCEx+y}$ would better describe a diverse sampling than $\mathcal{L}_{PNCEx}$. Indeed, for both N→T̂ and T→N̂ mapping directions, the models trained on heterogeneous data had higher performance (Fig. 2A, B). Models fare better with $\mathcal{L}_{PNCEx+y}$ than $\mathcal{L}_{PNCEx}$, either with or without heterogeneous training, suggesting $\mathcal{L}_{PNCEx+y}$ promotes more optimal translation even with data imbalance.

(iii) Next, we evaluated the role of int and external negative patches on training. For N→Т̂ and T→N̂ translation, inclusion of both internal and external negatives in calculating $\mathcal{L}_{PNCEx+y}$ yielded faster convergence and a slight improvement in performance (Fig. 2B). Our findings support theoretical predictions that $\mathcal{L}_{PNCEx+y}$ along with more diverse negative sampling and training data can better capture the mutual information and heterogeneity of original data distributions and enhance translation.

(iv) Regularization may be addressed by tuning the temperature parameter τ [16]. Comparisons suggest that varying τ for $\mathcal{L}_{PNCEx+y}$ can modulate the differences in FID between the canonically and heterogeneously trained models (Fig. 2C). Hence, τ and other measures such as internal and external training may allow for the adjustment of the degree of fitting to promote minimal reconstruction error (for clinical translation purposes) or a more optimal degree of T/N mismatch for further biological inquiry.

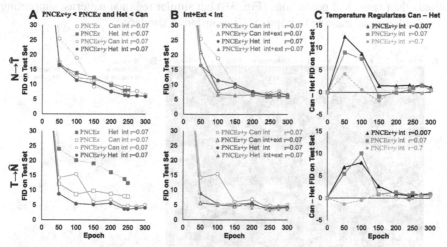

Fig. 2. Corroboration of theoretical predictions for N→Т̂ (top row) and T→N̂ (bottom row) by FID (lower score is better) over epochs. (**A**) Dual $\mathcal{L}_{PNCEx+y}$ outperforms single $\mathcal{L}_{PNCEx}$ and heterogeneous (het) models outperform canonical (can) models. (**B**) Use of internal and external (int+ext) negative patches in CL can improve performance over internal (int) negative patches alone. (**C**) Applying $\mathcal{L}_{PNCEx+y}$ with varying levels of temperature τ can tune FID differences between can and het models, suggesting that τ influences the degree of CL regularization.

4.2 T/N Translation Exhibits Both Accuracy and Residual T/N Mismatch

CUT models with $\mathcal{L}_{PNCEx+y}$ loss, int + ext negatives, and $\tau = 0.07$ yield realistic images with excellent anatomical agreement of PET uptake in cerebral/cerebellar gray/white matter, ventricles, orbits, skull, and background (Fig. 3A), performing well in patients with low and high T across AD severity. Local T/N relationships are preserved, with regions of high T generally corresponding to hypometabolism (N) and vice versa.

Yet, we also focus on the reconstruction error in the CUT output to validate spatial patterns from our original 6 T/N mismatch groups based on clustering on residuals of regressions of N vs. T data [3]. T/N canonical patients generally had the most comparable maps between observed and predicted maps. T/N resilient patients had more tau than expected given the level of hypometabolism ($T > \hat{T}$; higher observed T than expected $\hat{T}$) and less hypometabolism than expected given tau ($N < \hat{N}$; lower observed N than expected $\hat{N}$). Conversely, T/N susceptible patients had less tau than expected given degree of hypometabolism ($T < \hat{T}$; lower observed T than expected $\hat{T}$) and more hypometabolism than expected given tau ($N > \hat{N}$; higher observed N than expected $\hat{N}$).

Translation differences (residual = observed SUVR – expected SUVR) had more accurate spatial and directional patterns than original linear T/N groups (Fig. 3B, C); regions with limbic/cortical resilience or susceptibility showed lower residuals than original regression models (**Supplementary** Figs. 1, 2). Compared to N → $\hat{T}$, T→$\hat{N}$ maps had analogous yet more accurate output. As forecasted and confirmed with FID experiments (Fig. 2A), heterogeneous model predictions (Fig. 3C) had lower magnitude residuals than canonical predictions (Fig. 3B) but similar regional patterns, supporting the role of balanced training to improve CL translation similarity.

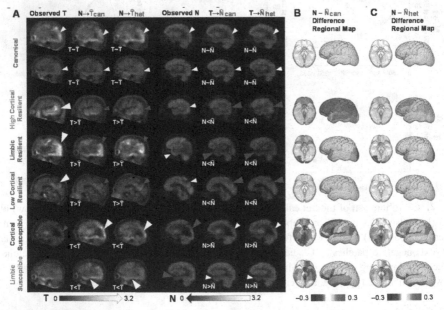

Fig. 3. Translation output. (**A**) Representative SUVR images for N→$\hat{T}$ (left) and T→$\hat{N}$ (right). Group T→$\hat{N}$ translation difference maps with (**B**) canonical and (**C**) heterogeneous residuals.

4.3 T/N Translation Mismatch May Reflect Non-AD Copathology Biomarkers

Heterogeneous T→$\hat{N}$ output had better correlation with observed copathology markers [22, 23] (Fig. 4C) than canonical output (Fig. 4B) and original input (Fig. 4A), implying

CUT learns some extent of "hidden" copathology patterns in N not "directly" seen on input T alone, but not as well as actual observed N used to compute copathology measures (Fig. 4D). Thus, CUT maps reflect both AD and non-AD factors.

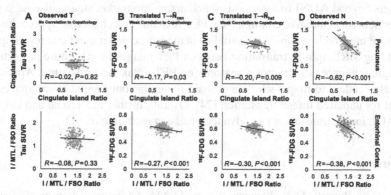

Fig. 4. Associations between observed data cingulate island ratio (top row) and inferior temporal gyrus/medial temporal lobe/frontal supraorbital gyrus (I/MTL/FSO) (bottom row) ^{18}F-FDG ratios with (**A**) observed T, (**B**) canonical T → $\hat{N}$, (**C**) heterogeneous T→$\hat{N}$ and (**D**) observed N in relevant regions. Spearman correlations, P values and regression lines are shown.

5 Conclusions

We report several theoretical and empirical findings from a study of CL divergences between latent distributions and the mapping between tau and ^{18}F-FDG PET images. Optimization of the sum of PNCE losses from different data domains acts through contrastive divergences between query data and each original data distribution [7, 8, 12]. *In turn, this multi-domain sum optimizes a cross entropy H (and thus KL divergence) between the query data distribution and the mixture of data distributions.*

(**i**) The multi-domain PNCE loss $\mathcal{L}_{PNCEx+y} = \mathcal{L}_{PNCEx} + \mathcal{L}_{PNCEy}$ compares query, positive and negative patches as a cross entropy H between the output and a mixture of data distributions in a shared latent space. This loss can outperform a single loss by maximizing mutual information between the output and both original data domains.
(**ii**) The multi-domain PNCE models benefit from diversified training implementation from heterogeneous patients compared to canonical patients with low residuals.
(**iii**) The multi-domain PNCE loss performs better with internal and external negative patches for negative sampling diversity than with internal negative patches alone.
(**iv**) The CL temperature τ may act as a form of adjust CL loss regularization.

Applying these results, we demonstrate that PNCE image translation converts between tau and ^{18}F-FDG PET for realistic, accurate output across regional anatomy and uptake mapping between T and N in the full AD spectrum. Indeed, areas of higher T demonstrated more severe hypometabolism on N, and vice versa. Yet, CUT still displayed some reconstruction error, highlighting how T/N groups had similar yet attenuated residuals compared to original regression residuals. Translated values correlated with copathology factors (of α-synuclein and TDP-43), indicating that CUT can learn and convey knowledge of non-AD copathologies within the AD mapping of T and N.

The present work has several limitations. Additional efforts can apply translation on datasets beyond ADNI to assess mismatch across biomarker status, disease severity and other neurodegenerative cohorts. New PET domains and clustering approaches may derive more mismatch groups. Since studies demonstrate better synthetic image quality when adding MRI input in translation between PET images [4, 6], future projects may disentangle influences of structural anatomy on PET uptake in translation. Application of several encoders to map each data domain to similar but unique latent spaces may be useful for unpaired image translation [24]. Overall, further theoretical and empirical study of CL image translation can advance precision medicine in AD.

References

1. Jack, C.R., Bennett, D.A., Blennow, K., et al.: NIA-AA research framework: toward a biological definition of Alzheimer's disease. Alzheimer's Dement. **14**, 535–562 (2018)
2. Das, S.R., Lyu, X., Duong, M.T., et al.: Tau-atrophy variability reveals phenotypic heterogeneity in Alzheimer's disease. Ann. Neurol. **90**, 751–762 (2021)
3. Duong, M.T., Das, S.R., Lyu, X., et al.: Dissociation of tau pathology and neuronal hypometabolism within the ATN framework of Alzheimer's disease. Nat. Commun. **13**(1), 1495 (2022)
4. Chen, K.T., Gong, E., Macruz, F.D.C., et al.: Ultra-Low-Dose ^{18}F-Florbetaben Amyloid PET Imaging Using Deep Learning with Multi-Contrast MRI Inputs. Radiology **290**, 649–656 (2019)
5. Ding, Y., Sohn, J.H., Kawczynski, M.G.: A deep learning model to predict a diagnosis of alzheimer disease by using ^{18}F-FDG PET of the brain. Radiology **290**(2), 456–464 (2019). https://doi.org/10.1148/radiol.2018180958
6. Zhou, B., Wang, R., Chen, M.-K., et al. Synthesizing multi-tracer PET images for alzheimer's disease patients using a 3D unified anatomy-aware cyclic adversarial network. In: Proceedings of the 34th International Conference on Medical Image Computing & Computer Assisted Intervention (2021)
7. Chopra, S., Hadsell, R., LeCun, Y.: Learning a similarity metric discriminatively, with application to face verification. In: Conference on Computer Vision and Pattern Recognition (2005)
8. Avd, O., Li, Y., Vinyals, O.: Representation Learning with Contrastive Predictive Coding. arXiv (2019)
9. Wang, T., Isola, P.: Understanding contrastive representation learning through alignment and uniformity on the hypersphere. In: Proceedings of the 37th International Conference in Machine Learning (2020)
10. Zimmerman, R.S., Sharma, Y., Schneider, S., et al.: Contrastive learning inverts the data generating process. In: Proceedings of the 38th International Conference in Machine Learning (2021)
11. Khosla, P., Teterwak, P., Wang, C., et al.: Supervised contrastive learning. In: Proceedings of the 34th Conference on Neural Information Processing Systems (2020)
12. Park, T., Efros, A.A., Zhang, R., Zhu, J.-Y.: Contrastive learning for unpaired image-to-image translation. In: European Conference on Computer Vision (2020)

13. Liu, M.-Y., Breuel, T., Kautz, J.: Unsupervised image-to-image translation networks. In: Proceedings of the 31st Conference on Neural Information Processing Systems (2017)
14. Jiang, Z., Chen, T., Chen, T., Wang, Z.: Improving contrastive learning on imbalanced seed data via open-world sampling. In: Proceedings of the 34th International Conference on Medical Image Computing & Computer Assisted Intervention (2021)
15. Chen, H., Zhao, L., Wang, Z., et al.: Artistic style transfer with internal-external learning and contrastive learning. In: Proceedings of the 34th International Conference on Medical Image Computing & Computer Assisted Intervention (2021)
16. Robinson, J., Sun, L., Yu, K., et al.: Can contrastive learning avoid shortcut solutions? Proceedings of the 35th Conference on Neural Information Processing Systems (2021)
17. Tian, Y., Krishnan, D., Isola, P.: Contrastive multiview coding. In: Proceedings of the 37th International Conference in Machine Learning (2020a)
18. Tian, Y., Sun, C., Poole, B., Krishnan, D., Scmid, C., Isola, P.: What makes for good views for contrastive learning? Proceedings of the 34th Conference on Neural Information Processing Systems (2020b)
19. Wen, Z., Li, Y.: Toward understanding the feature learning process of self-supervised contrastive learning. In: Proceedings of the 38th International Conference in Machine Learning (2021)
20. Skandarani, Y., Jodoin, P.-M., Lalande, A.: GANs for Medical Image Synthesis: An Empirical Study. arXiv (2021)
21. Haarburger, C., Horst, N., Truhn, D., et al.: Multiparametric magnetic resonance image synthesis using generative adversarial networks. Lawonn, Raidou R.G. (eds.) Eurographics Workshop on Visual Computing for Biology and Medicine (2019)
22. Buciuc, M., Botha, H., Murray, M.E., et al.: Utility of FDG-PET in diagnosis of Alzheimer-related TDP-43 proteinopathy. Neurology **85**, e23–e34 (2020)
23. Patterson, L., Firbank, M.J., Colloby, S.J., et al.: Neuropathological changes in dementia with lewy bodies and the cingulate island sign. J. Neuropathol. Exp. Neurol. **78**(8), 714–724 (2019)
24. Han, J., Shoeiby, M., Petersson, L., Armin, M.A.: Dual contrastive learning for unsupervised image-to-image translation. In: Conference on Computer Vision and Pattern Recognition. IEEE (2021)

Multi-shell dMRI Estimation from Single-Shell Data via Deep Learning

Reagan Dugan[1,2]([✉]) [iD] and Owen Carmichael[2] [iD]

[1] Louisiana State University, Baton Rouge, LA, USA
[2] Pennington Biomedical Research Center, Baton Rouge, LA, USA
reagan.dugan@pbrc.edu

Abstract. Diffusion magnetic resonance imaging (dMRI) data acquired with multiple diffusion gradient directions and multiple b-values ("multi-shell" data) enables compartmental modeling of brain tissues as well as enhanced estimation of white matter fiber orientations via the orientation distribution function (ODF). However, multi-shell dMRI acquisitions are time consuming, expensive and difficult in certain clinical populations. We present a method to estimate high b-value volumes from low b-value volumes via deep learning. A 3-dimensional U-NET architecture is trained from multi-shell dMRI training data to synthesize a high b-value volume from a diffusion gradient direction, given a low b-value volume from that same gradient direction as input. We show that our method accurately synthesizes high b-value (2000 and 3000 s/mm^2) volumes from low b-value (1000 s/mm^2) input volumes when applied to simulated and real, public-domain human dMRI data. We also show that synthesized multi-shell dMRI data gives rise to accurate compartmental model parameters and ODFs. Finally we demonstrate good out-of-training-sample generalization to previously-unseen diffusion gradient directions and different MRI scanners. Deep learning based estimation of high b-value dMRI volumes has the potential to combine with pulse sequence accelerations to enhance time efficiency of multi-shell dMRI protocols.

Keywords: Diffusion MRI · Multi-shell · Deep learning

1 Introduction

Diffusion magnetic resonance imaging (dMRI) is a non-invasive imaging technique that utilizes the random motion of water molecules in the brain to probe white matter microstructural properties. This is accomplished by collecting multiple image volumes with differing diffusion gradients to sensitize the signal to diffusion along differing spatial directions. A dMRI acquisition in which the MR signal is diffusion-sensitized to differing degrees across volumes (a multi b-value or multi-shell acquisition) can be useful for enhancing estimation of local tissue microstructure and/or white matter fiber directions, while reducing the partial volume effects that limit single-shell dMRI [1, 2]. Due to these advantages, multi-shell dMRI has been deployed to study brain maturation, aging, and other phenomena [2–5].

A key limitation of multi-shell dMRI is its long scan times. For example, a standard multi-shell protocol similar to a published one [5], using standard 3 T technology (b = 1000, 2000, and 3000 s/mm^2; 40 directions per shell; GE MR750W magnet) takes about 30 min; other published ones are as long as 50 min [1, 6]. Many individuals with neurological or orthopedic conditions find this duration intolerable; adding structural or functional sequences put these populations further out of reach. Long scan times make multi-shell dMRI expensive to acquire, as well. For these reasons, minimization of the total time of multi-shell dMRI protocols is a high priority.

We propose to reduce the total time of multi-shell dMRI protocols through machine learning based data augmentation. We use a convolutional neural network (CNN) to predict the appearance of dMRI volumes at higher b-values based on the appearance of volumes that were collected at lower b-values. The CNN is trained from a set of dMRI acquisitions that include both lower and higher b-value volumes. Similar work fit a spherical harmonics (SH) representation to the low b-value data, and estimated high b-value SH coefficients from the low b-value coefficients via machine learning [7, 8], but this approach is hindered by the limitations of SH, including its limited ability to accurately model white matter fiber orientations [9], its unreliable estimates in non-white matter tissues [10], and limited applicability to tissue compartmental modeling. Another method used a CNN to estimate white matter compartmental model parameters based on sparse multi-shell dMRI data [11], but like the SH method, this approach lacks flexibility, specifically to apply synthesized data to ODF estimation. In contrast, our synthesized multi-shell dMRI data sets can be applied to any ODF estimator, compartmental modeler, or other dMRI application that may arise. Other related work used machine learning to estimate missing dMRI slices—rather than missing dMRI shells-- given a dMRI data set with partial brain coverage [12]. Finally, a variety of methods have been presented for reducing the per-volume time of dMRI acquisition through pulse sequence acceleration [13, 14]. These methods require novel pulse sequences that are not currently universally available, but when they do become available they can be used seamlessly with our dMRI synthesis technique to achieve even shorter protocol times.

Section 2 presents our neural network architecture that takes a low b-value dMRI volume as input and synthesizes a corresponding volume at a higher b-value. Section 3 presents our tests of the method on a public-domain multi-shell real human dMRI dataset and on simulated multi-shell data. This section also presents our out-of-sample generalization experiments on data from differing MRI machines and differing pulse sequences. Network performance was evaluated in terms of ability to recreate exact voxel values, ability to generate similar tissue characteristic estimates, and ability to generate similar diffusion ODFs compared to ground truth.

2 Methods

Overview (Fig. 1). The proposed approach utilizes a 3D CNN to predict high b-value volumes from low b-value volumes. The network is trained on a set of diffusion weighted volume pairs, with the two volumes in a pair having approximately equal diffusion gradient directions but different b-values. A low b-value volume is the input to the network and convolutional kernels transform it into the corresponding high b-value volume; during training, kernel parameters are tuned to achieve this transformation of the training set. Once the network is trained, it is given a novel low b-value volume as input, and a predicted high b-value output volume is produced as output. The low b-value input volumes and high b-value output volumes are then used together to perform multi-shell tissue compartment modeling or ODF estimation.

Network Architecture. The CNN utilizes a U-NET architecture. This U-NET includes a contracting half that utilizes 2 3D convolutional layers with $3 \times 3 \times 3$ kernels followed by a 3D max pooling layer to reduce the size of an input volume by a factor of 2. This is repeated until the $96 \times 96 \times 96$ input volume is reduced to a $6 \times 6 \times 6$ feature map. The contracting half is paired with an expanding half that includes a 3D convolution transpose layer to upsample the volume by a factor of 2 followed by 2 3D convolution layers with 3x3x3 kernels. These layers are repeated until the volume has reached the original input shape, and a final 3D convolution layer with one filter is used to produce the final output volume [15]. The number of filters used in each convolution layer started at 16 in the first step and increased by a factor of 2 after every dimension reduction to a max at 25 in the lowest dimension layers and then decreased after every upsampling until returning to the original 16 filters. The exact filter progression can be found in Fig. 1. Every convolution layer used a ReLU activation function. The structure can be seen in Fig. 1. The network was trained using an 80%:10%:10% train:test:evaluate split and a batch size of 4 which was the highest value allowed by system memory. Training was performed for 50 epochs using a mean absolute error loss function with the Adam optimizer and a learning rate of 0.0001. The number of epochs was chosen after testing various values and choosing the value that produced the lowest evaluation loss without overfitting the model. A model was deemed to be overfitting if the validation loss plateaued while training loss continued to decrease. The network was made using Tensorflow 2.0 with the Keras API. Training datasets were stored in a TFRecord to improve training efficiency and minimize memory footprint and contained a total of 4766 training examples. All computing was performed on a system with a Quadro 5000 GPU which took approximately 90 min for network training. All low b-value volumes were divided by the scan's mean $b = 0$ volume to remove intensity variations due to T2-weighting and RF inhomogeneity. This also scaled the volume intensities to the range [0,1] which is ideal for stability of the neural network during training.

Simulated Multi-shell Data from In-house Single-Shell Data: Preprocessing. An in-house, single b-value data set was augmented via simulation of higher b-value images with the corresponding anatomy to arrive at a simulated multi-shell dataset. Single-shell dMRI scans from 97 participants in the Bogalusa Heart Study [16] were collected on a GE 3.0T Discovery MR750w scanner (4 $b = 0$ volumes, 40 directional volumes at $b = 1000$ s/mm^2). The `eddy_correct` tool (FSL 5.0) was used to correct for eddy current distortions and head movement. A diffusion tensor was then fit at each voxel using `DTIFIT` (FSL 5.0). White matter probability maps were generated using SPM12 [17].

Simulated Multi-shell Data from In-house Single-shell Data: Data Generation. These images were used to simulate corresponding volumes at b $= 2000$ and b $= 3000$ s/mm^2 using the `single_tensor` function from Diffusion

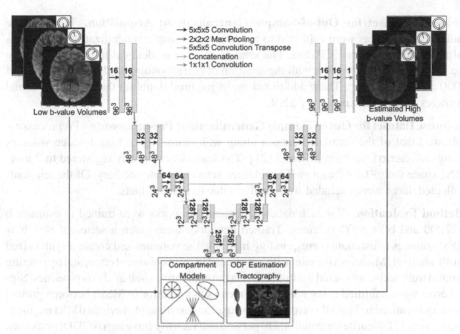

Fig. 1. Overview diagram of proposed method. Neural network uses a U-NET architecture with a contracting half that learns feature maps and an expanding half that produces a new volume from the feature maps. The input and output of the network is individual volumes. All low b-value volumes are fed to the network to produce a set of high b-value volumes. The dataset is then combined for multi-shell compartmental tissue modeling and ODF analyses.

Imaging in Python [18]. At each voxel diffusion tensor eigenvalues, eigenvectors and the b = 0 signal were used to simulate the signal at specified high b-values and gradient directions. A randomly-generated gradient direction discrepancy between 0 and 5° in magnitude was added to the high b-value gradient direction to match the low vs. high b-value direction discrepancies seen in the Human Connectome Project data set.

Human Connectome Project: Data. 267 multi-shell dMRI scans from the Human Connectome Project (HCP) were downloaded. Preprocessed data from healthy young adults had b-values of 1000, 2000, and 3000 s/mm^2, 90 directions per shell, and 18 b = 0 volumes [19]. HCP provided the large multi-shell data set required for proof-of-concept tests of the real-world use case of machine learning based high b-value synthesis.

Human Connectome Project: Pre-processing. All HCP scans were registered to 2 mm^3 MNI space using the FSL FLIRT tool to match the in-house scans described below [20]. Because gradient directions at b = 1000 did not match perfectly with those at b = 2000 or b = 3000, each b = 1000 direction was matched to its most similar b = 2000 and b = 3000 direction up to a tolerance of 5°. This data set was partitioned into train and test sets to evaluate the method.

In-house Dataset for Out-of-Sample Generalization: Acquisition. Ten in-house multi-shell data sets were collected to assess out-of-sample generalization of the neural network trained on HCP data. The scan protocol was identical to that described in the simulated dataset section with the addition of corresponding shells with b-values of 2000 and 3000 s/mm^2. These additional shells required doubling the number of signal averages to achieve satisfactory SNR.

In-house Dataset for Out-of-Sample Generalization: Pre-processing. Preprocessing followed that of the simulated dataset along with denoising of high b-value volumes using `dwidenoise` from MRtrix3 [21]. The scans were linearly registered to 2 mm^3 MNI space with `FLIRT` using a T1-weighted scan as an intermediary. Of the ten scans collected, three were excluded from analysis due to scan artifacts.

Method Evaluation. For each data set, separate networks were trained to estimate b = 2000 and b = 3000 volumes. Trained networks were given a series of new b = 1000 volumes to estimate corresponding high b-value volumes and create a synthesized multi-shell set. Model performance was evaluated in terms of signal error, and by running ground truth and synthesized multi-shell data through multi-shell analysis pipelines. Signal error was quantified using normalized mean squared error (NMSE) between ground truth and synthesized signal vectors at each voxel as was done in previous dMRI augmentation work [7]. Neurite Orientation Dispersion and Density Imaging (NODDI) was used to estimate mean white matter values of neurite density index (NDI), orientation dispersion index (ODI), and free water fraction [1] in ground-truth and synthesized data sets. Similarity in NODDI parameters between synthesized and ground truth data was quantified via two-way, mixed, single score intraclass correlation coefficient (ICC). ODFs were computed using multi-tissue constrained spherical deconvolution [10]. Ground truth and synthesized white matter ODFs were compared using the angular correlation coefficient [22]. For the simulated and HCP data tests, 10 scans were excluded from the training set and used for method evaluation.

Comparison to SH Methods. We compared the proposed method that synthesizes high b-value raw signal intensity values to a corresponding method that synthesizes high b-value SH coefficients as in prior work [7, 8]. The model was a 2D version of the proposed U-NET where all 3D layers were replaced with their 2D equivalents. The input was the 45 SH coefficients of each voxel in a slice from b = 1000 data, and the output was the 45 SH coefficients from the same slice at b = 2000. SH coefficients were computed using the same method as in the work by Jha et al. [8]. The model was trained to compute b = 2000 SH coefficients from b = 1000 SH coefficients using SH representations of the HCP dataset and following the same training strategy described in previous sections. This modified model was compared with the proposed method through the white matter NMSE metric described in the previous section.

3   Results

Simulated Data. Signal differences between ground truth and synthesized b = 2000 and b = 3000 images were low with NMSE of 0.27% and 1.3% respectively (Table 1). NODDI parameter agreement between ground truth and synthesized was uniformly high

(ICC > 0.97, Table 1). The ACC between the ground truth and estimated ODFs across subjects was also high (0.663 ± 0.008, mean ± standard error).

Table 1. Results of experiments on all datasets. ICC values are reported with 95% confidence interval in square brackets, signal NMSE and ACC reported as mean across all evaluation subjects with standard error.

	NDI ICC	ODI ICC	FWF ICC	b = 2000 NMSE	b = 3000 NMSE	Mean ACC
Simulated Dataset	0.99 [0.99, 0.99]	0.97 [0.91, 0.99]	0.98 [0.93, 0.99]	0.27% ± 0.02%	1.32% ± 0.19%	0.66 ± 0.01
HCP Dataset	0.99 [0.99, 0.99]	0.98 [0.95, 0.99]	0.99 [0.96, 0.99]	1.05% ± 0.06%	2.04% ± 0.06%	0.70 ± 0.01
Out-of-sample Dataset	0.87 [0.43, 0.97]	0.90 [0.54, 0.98]	0.91 [0.60, 0.98]	2.25% ± 0.21%	6.06% ± 0.61%	0.68 ± 0.01
DNN [7]				3.49%	5.63%	
MSR-Net [8]				1.59%		
Proposed Method with SH				4.62% ± 0.56%		

Human Connectome Project. Signal differences between ground truth and synthesized b = 2000 and b = 3000 images were uniformly low with NMSE of 1.05% and 2.04% respectively (Table. 1). The b = 2000 results were superior to two competing methods using the same evaluation metric and superior to the method proposed by Koppers et al. at b = 3000 (Table 1). Agreement in NODDI parameters between ground truth and synthetic data was uniformly high (ICC > .98). Mean ACC between the ground truth and estimated ODFs across subjects was 0.705 ± 0.006 (mean ± standard error).

Out-of-Training-Sample Generalization. Signal differences between ground truth and synthesized b = 2000 and b = 3000 images remained relatively low with NMSE values of 2.25% and 6.06% respectively (Table 1). The larger error in b = 3000 estimation is likely due to the large increase in noise in this dataset compared to that from HCP. Agreement between ground truth and synthetic data based NODDI parameter estimates was high for NDI, ODI, and FWF (ICC point estimates of 0.87, 0.9, and 0.91). ACC between ground truth and estimated ODFs was high: 0.689 ± 0.002 (mean ± standard error). Figure 2 shows a comparison of ground truth and synthesized volumes for the b = 2000 and b = 3000 images along with NODDI parameter maps. The leftmost column shows the ground truth data, middle column the synthesized data from the trained model, and the right column shows a difference image between ground truth and synthesized. The difference images show that the model performs well at synthesizing b = 2000 and b = 3000 volumes with slightly lower performance at b = 3000 likely caused by the

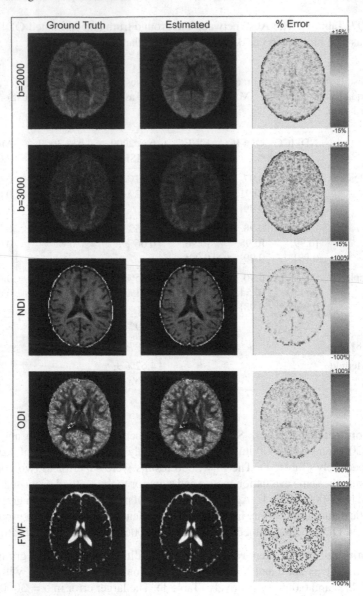

Fig. 2. Comparison of example slices from ground truth and synthesized volumes for b = 2000 and b = 3000, along with NODDI parameter maps derived from those volumes. Data is from the out-of-sample dataset. Error maps represent the percent error with negative error in blue and positive error in red. (Color figure online)

noisy ground truth data. The NDI and ODI difference images show that the majority of high error voxels are at the edges of the brain or in the ventricles. For FWF, the error outside the ventricles has no clear pattern and appears to be noise removal.

Comparison to Competing Methods. The method that synthesizes SH coefficients [8] generated higher-error NMSE (1.59%) on their own real data, using their own implementation, compared to our implementation of a raw signal synthesizer applied to simulated (0.27%) and HCP data (1.06%). Our attempt to apply a similar SH synthesizer to HCP data produced even higher error (4.62%), suggesting that the higher error of [8] was not simply due to the data set being more challenging. The DNN method similarly produced higher error than our approach (3.49%) [7]. Together these results suggest that our raw signal synthesizing method is competitive with pre-existing methods.

4 Conclusion

In this work we presented a method that estimates entire high b-value dMRI image volumes from low b-value volumes via a deep learning model. Our experiments suggest that the method synthesizes raw data that is highly similar to ground truth, and that raw data produces ODFs and tissue compartmental models that are also highly similar to ground truth. This method provides superior data synthesis compared to a competing SH. A potential next step would be to use the network to generate a high b-value volume at a different gradient direction, with that novel gradient direction provided as an input to the network. Copies of the code and model generated for this work can be found at https://github.com/rdugan3/MS_DTI_MLCN_2023.

References

1. Zhang, H., Schneider, T., Wheeler-Kingshott, C.A., Alexander, D.C.: NODDI: practical in vivo neurite orientation dispersion and density imaging of the human brain. Neuroimage **61**, 1000–1016 (2012). https://doi.org/10.1016/j.neuroimage.2012.03.072
2. Jeurissen, B., Leemans, A., Jones, D.K., Tournier, J.D., Sijbers, J.: Probabilistic fiber tracking using the residual bootstrap with constrained spherical deconvolution. Hum. Brain Mapp. **32**, 461–479 (2011). https://doi.org/10.1002/hbm.21032
3. Kunz, N., et al.: Assessing white matter microstructure of the newborn with multi-shell diffusion MRI and biophysical compartment models. Neuroimage **96**, 288–299 (2014). https://doi.org/10.1016/j.neuroimage.2014.03.057
4. Chang, Y.S., et al.: White matter changes of neurite density and fiber orientation dispersion during human brain maturation. PLoS One **10**, (2015). https://doi.org/10.1371/journal.pone.0123656
5. Nazeri, A., et al.: Functional consequences of neurite orientation dispersion and density in humans across the adult lifespan. J. Neurosci. **35**, 1753–1762 (2015). https://doi.org/10.1523/JNEUROSCI.3979-14.2015
6. Tournier, J.D., Calamante, F., Connelly, A.: Determination of the appropriate b value and number of gradient directions for high-angular-resolution diffusion-weighted imaging. NMR Biomed. **26**, 1775–1786 (2013). https://doi.org/10.1002/nbm.3017
7. Koppers, S., Haarburger, C., Merhof, D.: Diffusion MRI Signal Augmentation: From Single Shell to Multi Shell with Deep Learning (2016)
8. Jha, R.R., Nigam, A., Bhavsar, A., Pathak, S.K., Schneider, W., Rathish, K.: Multi-shell D-MRI reconstruction via residual learning utilizing encoder-decoder network with attention (MSR-Net). In: Proceedings of the Annual International Conference of the IEEE Engineering in Medicine and Biology Society. EMBS. 2020-July, pp. 1709–1713 (2020). https://doi.org/10.1109/EMBC44109.2020.9175455

9. Yan, H., Carmichael, O., Paul, D., Peng, J.: Estimating fiber orientation distribution from diffusion MRI with spherical needlets. Med. Image Anal. **46**, 57–72 (2018). https://doi.org/10.1016/j.media.2018.01.003
10. Jeurissen, B., Tournier, J.D., Dhollander, T., Connelly, A., Sijbers, J.: Multi-tissue constrained spherical deconvolution for improved analysis of multi-shell diffusion MRI data. Neuroimage **103**, 411–426 (2014). https://doi.org/10.1016/j.neuroimage.2014.07.061
11. Gibbons, E.K., et al.: Simultaneous NODDI and GFA parameter map generation from sub-sampled q-space imaging using deep learning. Magn. Reson. Med. **81**, 2399–2411 (2019). https://doi.org/10.1002/mrm.27568
12. Hong, Y., Chen, G., Yap, P.-T., Shen, D.: Reconstructing high-quality diffusion MRI data from orthogonal slice-undersampled data using graph convolutional neural networks. In: Shen, D., et al. (eds.) Medical Image Computing and Computer Assisted Intervention – MICCAI 2019, pp. 529–537. Springer International Publishing, Cham (2019)
13. Feinberg, D.A., et al.: Multiplexed echo planar imaging for sub-second whole brain FMRI and fast diffusion imaging. PLoS One **5** (2010). https://doi.org/10.1371/journal.pone.0015710
14. Setsompop, K., et al.: Improving diffusion MRI using simultaneous multi-slice echo planar imaging. Neuroimage **63**, 569–580 (2012). https://doi.org/10.1016/j.neuroimage.2012.06.033
15. Ronneberger, O., Fischer, P., Brox, T.: U-Net: convolutional networks for biomedical image segmentation. In: Navab, N., Hornegger, J., Wells, W.M., Frangi, A.F. (eds.) Medical Image Computing and Computer-Assisted Intervention – MICCAI 2015, pp. 234–241. Springer International Publishing, Cham (2015)
16. Berenson, G.S.: Bogalusa heart study: a long-term community study of a rural biracial (Black/White) population. Am. J. Med. Sci. **322**, 267–274 (2001). https://doi.org/10.1097/00000441-200111000-00007
17. Ashburner, J., Friston, K.J.: Unified segmentation. Neuroimage **26**, 839–851 (2005). https://doi.org/10.1016/j.neuroimage.2005.02.018
18. Descoteaux, M.: High Angular Resolution Diffusion MRI: from Local Estimation to Segmentation and Tractography (2008)
19. Glasser, M.F., et al.: The minimal preprocessing pipelines for the human connectome project and for the WU-Minn HCP consortium. Neuroimage **80**, 105–12404 (2013). https://doi.org/10.1016/j.neuroimage.2013.04.127.The
20. Jenkinson, M., Bannister, P., Brady, M., Smith, S.: Improved optimization for the robust and accurate linear registration and motion correction of brain images. Neuroimage **17**, 825–841 (2002)
21. Cordero-Grande, L., Christiaens, D., Hutter, J., Price, A.N., Hajnal, J.V.: Complex diffusion-weighted image estimation via matrix recovery under genera noise models. Neuroimage **200**, 391–404 (2019)
22. Anderson, A.W.: Measurement of fiber orientation distributions using high angular resolution diffusion imaging. Magn. Reson. Med. **54**, 1194–1206 (2005). https://doi.org/10.1002/mrm.20667

A Three-Player GAN
for Super-Resolution in Magnetic
Resonance Imaging

Qi Wang[1]([✉])(iD), Lucas Mahler[1], Julius Steiglechner[1,2], Florian Birk[1,2](iD),
Klaus Scheffler[1,2](iD), and Gabriele Lohmann[1,2](iD)

[1] Max Planck Institute for Biological Cybernetics, Tübingen, Germany
qi.wang@tuebingen.mpg.de
[2] University Hospital of Tübingen, Tübingen, Germany

Abstract. Learning based single image super resolution (SISR) is well
investigated in 2D images. However, SISR for 3D magnetic resonance
images (MRI) is more challenging compared to 2D, mainly due to the
increased number of neural network parameters, the larger memory
requirement, and the limited amount of available training data. Current SISR methods for 3D volumetric images are based on generative
adversarial networks (GANs), especially Wasserstein GANs. Other common architectures in the 2D domain, e.g. transformer models, require
large amounts of training data and are therefore not suitable for the
limited 3D data. As a popular GAN variant, Wasserstein GANs sometimes produce blurry results as they may not converge to a global (or
local) optimum. In this paper, we propose a new method for 3D SISR
using a GAN framework. Our approach incorporates instance noise to
improve the training stability of the GAN, as well as a relativistic GAN
loss function and an updating feature extractor as the third player during the training process. Our experiments demonstrate that our method
produces highly accurate results using very few training samples. Specifically, we show that we need less than 30 training samples, which is a
significant reduction compared to the thousands typically required in
previous studies. Moreover, our method improves out-of-sample results,
demonstrating its effectiveness. The code is available in https://github.
com/wqlevi/threeplayerGANSR.

Keywords: Super-resolution · Brain MRI · GAN

1 Introduction

High-resolution MR images are crucial for obtaining a detailed view of anatomical structures and are essential for downstream MRI analysis. However, the
acquisition of high-resolution (HR) MRI is labor-intensive and susceptible to
motion artifacts during the scanning process. Furthermore, reducing scan time

Supplementary Information The online version contains supplementary material
available at https://doi.org/10.1007/978-3-031-44858-4_3.

A. Abdulkadir et al. (Eds.): MLCN 2023, LNCS 14312, pp. 23–33, 2023.
https://doi.org/10.1007/978-3-031-44858-4_3

often results in lower spatial resolution and the loss of fine details in the image structure. Therefore, much research in deep learning has emerged towards the single image super resolution (SISR) task, which aims to recover high-resolution MR images from lower-resolution images of the same subject.

As a result of the curse of dimensionality and limited availability of data, training a SISR model on 3D medical images is more challenging than training on common 2D images. Medical images, such as brain MRI, contain anatomical information in all three dimensions, making the task of SISR on 3D volumetric MRI the focus of this paper.

Despite its effectiveness, GAN training is known to be very unstable, sensitive to parameter changes, and requires careful design of the architecture, especially in the high dimension space. Wasserstein GAN (WGAN) [15], together with WGANGP [4], is one of the most effective and popular methods in training SR models in the 3D MRI domain. Despite their improvement from standard GAN, we have observed that the results from the WGAN (or WGANGP) models, e.g. DCSRN [2], are often prone to produce blurring results, which can be a consequence of their convergence to sub-optima and oscillating dynamics around the global optimum [16,27]. In this paper, we propose to tackle this problem by introducing instance noise to the model and incorporating a relativistic GAN loss [7] into a three-player adversarial training [11] framework.

Instance noise was originally implemented in [21] to balance the GAN training for 2D super resolution. Relativistic GAN was proposed by [7] as an efficient alternative to standard GAN loss functions, this was later implemented in [25] for the 2D SR task. The three-player game was studied in [1] and was implmented for improving classification tasks by [24]. This combination exhibits stable training dynamics and produces the *state-of-the-art* results, demonstrating superior generalization performance on previously unseen data compared to other models. Usually SR models are trained on datasets of thousand subjects, as in [2,26]. We observed that training on *dozens* of samples enables our model to achieve superior performance and generalizability, even on out-of-distribution (OOD) data. To summarise, our contributions in this work are the following:

- We show empirically that our model exhibits better convergence than other GAN models on 3D brain MRI data.
- We show that the performance of our model remains robust even with a limited number of training samples.
- We show our model can generalize well to datasets of different modalities and resolutions.

2   Related Work

Super-Resolution in MR Images. Learning based SISR has been actively studied in both 2D and 3D images. In a SISR task, the neural network aims to learn a non-linear mapping from the low resolution (LR) image to its high resolution (HR) reference. This is particularly suited to the learning capability

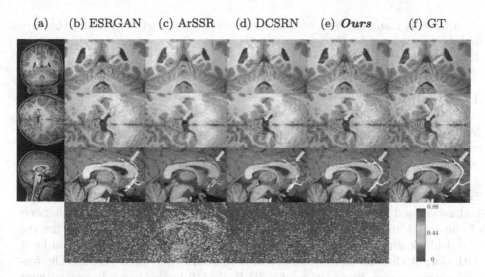

Fig. 1. Qualitative comparison on an in-sample subject using different models with a ×2 resolution upscale. *Top* three rows show MRI images from different models, the *last* row shows corresponding residual maps between SR images and the ground truth (GT), of the sagittal view. **(a)** shows the overview of whole brain MRI GT, with the yellow boxes indicating the zoomed-in region; The zoomed-in views of various SR results are shwon in **(b)-(e)** and the GT in **(f)**. Our results show the best visual quality, the most detailed recovery of the GT, and the smallest residuals in brain structure. Distinguishable differences are marked by *yellow* arrows. (Color figure online)

of convolutional neural networks (CNNs), as demonstrated in previous studies [3,10,25,30]. In 2D MR images, a squeeze-excitation attention network achieved remarkable results [29], while a transformer architecture achieved superior quality in a multiscale network [12]. However, since downstream analysis typically requires 3D volumes of MR images, stacking 2D slices of SR results can lead to artefacts. Therefore, implementing a 3D model can directly outperform 2D models [18].

This paper will focus only on 3D MR image training. Most of the novel architectures have not been adapted to 3D MRI data, due to the high dimensionality and lack of data. Thus, especially for the 3D MRI domain, the GAN category remains the mainstream training method, given its efficiency and high performance. An initial work by [18] discussed the implementation of CNN models on 3D MRI data for SISR tasks, demonstrating the advanced SR results produced by neural networks over traditional interpolation methods. In [2], the authors combined WGAN training with densely connected residual blocks and reported prominent image quality. While in [26], the authors used implicit neural representation to realise SR by generating images at an arbitrary scale factor. In [22], the authors proposed to use image gradients as a knowledge prior, for a better matching to local image patterns.

Although many of these studies successfully produced SR results, they are mainly constrained to the same modality or imaging sequence as the training dataset. Also the quality of the reconstruction is sometimes limited.

3   Methods

Model Design. Figure 2a shows an overview of our GAN model during training, consisting of a generator, a discriminator, and a feature extractor. The generator is comprised of 3 residual-in-residual blocks (RRDBs) [6] and a pixel-shuffle layer [20] for upsampling the feature maps. The discriminator contains four blocks comprised of strided convolution layers followed by instance normalization and Leaky ReLU activation functions. The numbers of the channels of the input and output are the same for neighboring blocks, except for the initial block which has to match the input image channel. A convolution layer with output channel equals one is added to restrict the final output. The feature extractor uses lower layers of a 3D ResNet10 before linear layers, without pretrained weights. All three networks are trained concurrently, where Gaussian noise is added to both real and generated inputs of the discriminator. The scheduled noise annealing ensures training stability and convergence to the optimal point, see Fig. 2b.

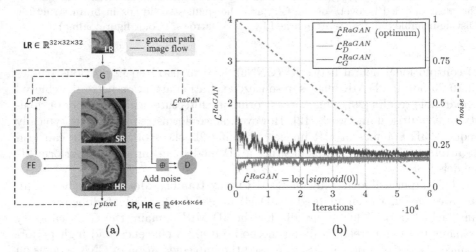

(a)   (b)

Fig. 2. (a) Schematic diagram of our network during training. *D, G, FE* denote *discriminator, generator* and *feature extractor* respectively. Low resolution volumes (LR) are fed to the generator to produce super-resolution images(SR), the high resolution images (HR) and SR are distinguished by the discriminator. (b) Convergence of the adversarial losses during training, see its result in Fig. 1e. Both $\mathcal{L}_D^{RaGAN}$ and $\mathcal{L}_G^{RaGAN}$ converge to their theoretical optimum (noted as $\hat{\mathcal{L}}^{RaGAN}$ in green line), under the annealing noise variance (σ_{noise}). (Color figure online)

The generator takes LR images as input and outputs SR images, while the discriminator tries to distinguish between the generated SR images and the ground truth HR images (see Fig. 2a).

As the building block of the generator, we implemented three RRDBs [6,25]. Deep architectures can be prone to instability during training. Especially in a GAN model, gradients are prone to vanish or explode. Thus RRDBs were implemented to improve computational efficiency and performance while retaining training stability [13,25].

We add linearly annealed Gaussian noise [21] to both the SR and HR inputs of the discriminator during training, to avoid early phase divergence by forcing these two distribution overlapped. The Gaussian noise is from a standard normal distribution with its variance linearly decreasing per iteration from 1 to 0. This has been shown to be an effective and computationally cheap way of balancing the GAN dynamics and ensuring convergence [16,21]. As is shown in 2b, the adversarial losses of both the generator and discriminator converge to the optimal value (green line, details see supplementary material), when trained under annealed noise (gray dashed line).

Feature matching has been shown to improve the perceptual quality of images in previous studies [8,10]. In contrast to the standard perceptual loss which uses pretrained VGG19 network for feature extraction [10,13,25], we found that ResNet10, a shallower architecture, performed better in our experiments when updated without pretraining.

Objective Function. In order to generate images with both high perceptual quality and balanced training performance, we include a $\mathcal{L}_1$ loss in the image domain ($\mathcal{L}^{pixel}$), a perceptual loss ($\mathcal{L}^{perc}$), and a GAN loss as the objective function for the generator network. It is defined as:

$$\mathcal{L}_G = \mathcal{L}^{perc} + \alpha \mathcal{L}^{pixel} + \beta \mathcal{L}_G^{RaGAN} \tag{1}$$

where $\alpha = 0.01$ and $\beta = 0.005$ are as default in similar work [25]. The gradients of the generator are updated by the derivatives with respect to $\mathcal{L}_G$. The discriminator and the feature extracting network are updated based on their respective losses, $\mathcal{L}_D^{RaGAN}$ and $\mathcal{L}^{perc}$, as defined in the following.

Our perceptual loss is defined as a $\mathcal{L}_1$ loss in feature space, see Eq. 2, which simultaneously updates the feature extractor and generator. Here $I_{i,j,k}$ and $\hat{I}_{i,j,k}$ represent HR and SR intensity values at the i,j,k-th voxel in all three image dimensions (W, H, D), $\mathcal{F}(\cdot)$ represent the feature extractor network.

$$\mathcal{L}^{perc} = \frac{1}{W \times H \times D} \sum_{i,j,k=1}^{W,H,D} |\mathcal{F}(I_{i,j,k}) - \mathcal{F}(\hat{I}_{i,j,k})| \tag{2}$$

To ensure the efficiency and stability of the model, we use relativistic GAN loss [7] as our adversarial loss function. The critic network (the discriminator without the last sigmoid layer) is encouraged to distinguish real data as being more realistic than the average of the generated data, *vice versa* for the generator's GAN loss. The real samples are denoted as $x \sim \mathbb{P}$, with $\mathbb{P}$ the distribution of the real samples; the generated data are noted as $y \sim \mathbb{Q}$, with $\mathbb{Q}$ the distribution of the generated samples. We propose to use the relativistic averaged discriminator (D_{Ra}), where $C(\cdot)$ is the critic network.

$$D_{Ra}(x, y) = Sigmoid[C(x) - \mathbb{E}_{y \sim \mathbb{Q}} C(y)] \qquad (3)$$
$$D_{Ra}(y, x) = Sigmoid[C(y) - \mathbb{E}_{x \sim \mathbb{P}} C(x)] \qquad (4)$$

The relativistic GAN loss for the discriminator $(\mathcal{L}_D^{RaGAN})$ and the generator $(\mathcal{L}_G^{RaGAN})$ is defined below.

$$\mathcal{L}_D^{RaGAN}(x, y) = -\mathbb{E}_{x \sim \mathbb{P}}[\log(D_{Ra}(x, y))] - \mathbb{E}_{y \sim \mathbb{Q}}[\log(1 - D_{Ra}(x, y))] \qquad (5)$$
$$\mathcal{L}_G^{RaGAN}(x, y) = -\mathbb{E}_{x \sim \mathbb{P}}[\log(1 - D_{Ra}(y, x))] - \mathbb{E}_{y \sim \mathbb{Q}}[\log(D_{Ra}(y, x))] \qquad (6)$$

The optimal point is reached when real and fake samples are indistinguishable for the discriminator. In our context, the D_{Ra} decreases to 0 and the $\mathcal{L}^{RaGAN}$ converges to $-\log[sigmoid(0)]$.

4   Experimental Setup

Datasets. We used publicly available datasets from different scanners, imaging modalities or parts of body. Three of these datasets are brain MRI acquired at 3 T, which are part of the Human Connectome Project [23], the other one is a knee dataset with proton density (PD) contrast at 3 T from [19]. We name the datasets by the most distinguishable attribute of the each dataset, as described in the Table 1 below:

Table 1. All datasets used in the experiments

Dataset name	Dataset source	Resolution	Contrast	Number of subjects	Vendor
"Insample"	*Lifespan Pilot Project*(HCP [23])	0.8 mm	T1w	27	Siemens
"Contrast"	*Lifespan Pilot Project*(HCP [23])	0.8 mm	T2w	27	Siemens
"Resolution"	Young Adult study(HCP)	0.7 mm	T1w	1113	Siemens
"Knee"*	Fully sampled knees [19]	0.5/0.6 mm	PD	20	GE clinical

Implementation Details. During the training of the network, we simultaneously trained a generator, a critic, and a feature extractor network. To fit in the GPU memories, we crop the whole brain volumes into overlapping HR patches of size $64 \times 64 \times 64$. The LR patches were obtained by linearly down-sampling HR patches to a matrix size of $32 \times 32 \times 32$. The paired LR and HR patches were fed into the model for training.

All three networks are initialised by Kaiming initialization [5] and are optimised by Adam optimizers [9], using the default coefficients of $\beta_1 = 0.9$, $\beta_2 = 0.999$, and learning rate of 1×10^{-4}. The variance of the instance noise is scheduled to decrease linearly per epoch for 20 epochs, from $\sigma = 1$ to 0. The training is implemented using the PyTorch framework [17] and run for 60000 iterations, on an NVIDIA's Ampere 100 GPU.

Evaluation Metrics. To quantify SR quality, we use Peak-Signal-Noise-Ratio (PSNR), Structure-Similarity-Index-Measurement (SSIM), and Learned-Perceptual-Image-Patch-Similarity (LPIPS) [28]. However, compared with the other metrics, LPIPS better reflect image fidelity [14,28]. Here we use the 2D slice-wise LPIPS implementation provided by [28], which has been pretrained on large image datasets and are known to extract meaningful features that closely resemble human visual quality [14,28].

Table 2. Quantitative comparison of robustness among the *state-of-the-art* models on **Dataset "Insample"** and **Dataset "Resolution"** . Note the LPIPS score matches perceptual quality better than PSNR and SSIM, as shown in Fig. 1 and Fig. 3

Dataset names	Models	PSNR ↑	SSIM ↑	LPIPS↓
Dataset "Insample"	Tri-linear	33.038	0.876	0.084
	ESRGAN [25]	37.022	0.933	0.044
	DCSRN [2]	**37.635**	**0.954**	0.052
	ArSSR [26]	28.038	0.280	0.291
	ours	36.922	0.943	**0.037**
Dataset "Resolution"	ESRGAN	37.181	0.957	0.039
	DCSRN	**37.564**	0.962	0.051
	ArSSR	36.550	**0.970**	0.055
	ours	36.922	0.953	**0.038**

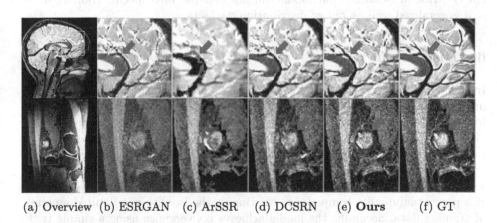

(a) Overview   (b) ESRGAN   (c) ArSSR   (d) DCSRN   (e) **Ours**   (f) GT

Fig. 3. Qualitative comparison on out-of-distribution datasets "Contrast" (*top row*) and "Knee" (*bottom row*) for different models. (a) shows the overall view of GT image and (b)–(f) are zoomed-in SR from corresponding methods; distinguishable differences are marked in *red* arrows (Color figure online)

5 Results

In-sample Super-Resolution Performance. For comparison, we trained all models, namely ESRGAN [25], ArSSR [26], DCSRN [2], and ours, on *Dataset "Insample"*. ESRGAN was only available in 2D model, so we reimplemented it in 3D (see supplementary for details).

The results are shown in Fig. 1. Note that our model recovered the ground truth with the most details, resulting the smallest residual map against GT. Especially with regard to small blood vessels and more detailed cerebellum structures (Fig. 1e). The quantitative results are shown in Table 2. Note that our method outperformed other models in LPIPS. Although DCSRN obtained the highest PNSR and SSIM, the LPIPS score reflects the perceptual quality and fine details of images best, as demonstrated in Fig. 1 and Table 2.

Zero-Shot Inference on OOD Data. To test the generalizability of our model, we use all models trained on the Dataset "Insample" and evaluated their performance on three different datasets: "Contrast", "Resolution", and "Knee". The ground truth of these OOD datasets exhibits significant variations in terms of contrast, resolution, and body parts, as are shown in Fig. 3.

Our model outperformed the other models in this experiment on all three datasets, by producing the best image details that most closely approximated the GT. Our model also exhibited superior image quality on the "Contrast" and "Knee" datasets, where it closely approximated the blood vessels and the round-shaped bone structure, as demonstrated and marked in *red arrows* in Fig. 3e. Our experiments showed that our approach can be generalized to different datasets with little decrease in performance. Notably, even when exposed to vastly different contrast or anatomy, our model did not collapse into specific training data and generalized well on out-of-distribution datasets.

6 Discussion

In this paper, an effective method is proposed for stabilizing GAN training, which outperforms *state-of-the-art* SR models on 3D brain MRI data. Our approach involves adding Gaussian noise to discriminator inputs, forming a three-player GAN game, and applying the relativistic GAN loss as the objective function. Our experimental results demonstrate that this method can ensure effective training convergence and achieve better generalizability. Several key findings are presented in this paper. First, the proposed model generates SR results with superior perceptual quality compared to existing methods, with very limited amount of training data. Secondly, the model achieves convergence using a simple trick of adding noise. Finally, the model exhibits strong generalizability and is less prone to overfitting, as evidenced by its successful performance on previously unseen datasets.

While the LPIPS score better reflects the perceptual quality of images, it is primarily designed for natural 2D images and may overlook information in the cross-plane of 3D MRI images. Therefore, a specialized metric for evaluating the perceptual quality of 3D MRI images is of high importance in the future.

Acknowledgements. This work is supported by the Deutsche Forschungsgemeinschaft (DFG - German Research Foundation), the grant numbers are DFG LO1728/2-1 and DFG HE 9297/1-1.

References

1. Balduzzi, D., Racaniere, S., Martens, J., Foerster, J., Tuyls, K., Graepel, T.: The mechanics of n-player differentiable games. In: Proceedings of the 35th International Conference on Machine Learning. Proceedings of Machine Learning Research, vol. 80, pp. 354–363. PMLR, 10–15 July 2018
2. Chen, Y., Xie, Y., Zhou, Z., Shi, F., Christodoulou, A.G., Li, D.: Brain MRI super resolution using 3d deep densely connected neural networks. In: 2018 IEEE 15th International Symposium on Biomedical Imaging (ISBI 2018). IEEE, April 2018. https://doi.org/10.1109/isbi.2018.8363679
3. Dong, C., Loy, C.C., He, K., Tang, X.: Learning a deep convolutional network for image super-resolution. In: Fleet, D., Pajdla, T., Schiele, B., Tuytelaars, T. (eds.) ECCV 2014. LNCS, vol. 8692, pp. 184–199. Springer, Cham (2014). https://doi.org/10.1007/978-3-319-10593-2_13
4. Gulrajani, I., Ahmed, F., Arjovsky, M., Dumoulin, V., Courville, A.C.: Improved training of Wasserstein GANs. In: Advances in Neural Information Processing Systems, vol. 30. Curran Associates, Inc. (2017)
5. He, K., Zhang, X., Ren, S., Sun, J.: Delving deep into rectifiers: surpassing human-level performance on imagenet classification. In: Proceedings of the IEEE International Conference on Computer Vision (ICCV), December 2015
6. Huang, G., Liu, Z., van der Maaten, L., Weinberger, K.: Densely Connected Convolutional Networks. In: Proceedings of the IEEE Conference on Computer Vision and Pattern Recognition (CVPR), July 2017
7. Jolicoeur-Martineau, A.: The relativistic discriminator: a key element missing from standard GAN. In: International Conference on Learning Representations (2019)
8. Justin, J., Alexandre, A., Li, F.: Perceptual losses for real-time style transfer and super-resolution (2016). https://doi.org/10.48550/ARXIV.1603.08155
9. Kingma, D.P., Ba, J.: Adam: a method for stochastic optimization. In: 3rd International Conference on Learning Representations, ICLR 2015, San Diego, CA, USA, May 7–9, 2015, Conference Track Proceedings (2015)
10. Ledig, C., et al.: Photo-realistic single image super-resolution using a generative adversarial network. CoRR abs/1609.04802 (2016)
11. Li, C., Xu, K., Zhu, J., Zhang, B.: Triple generative adversarial nets. In: Proceedings of the 31st International Conference on Neural Information Processing Systems, pp. 4091–4101. NIPS 2017, Curran Associates Inc. (2017)
12. Li, G., et al.: Transformer-empowered multi-scale contextual matching and aggregation for multi-contrast MRI super-resolution. In: 2022 IEEE/CVF Conference on Computer Vision and Pattern Recognition (CVPR), pp. 20604–20613 (2022). https://doi.org/10.1109/CVPR52688.2022.01998
13. Ma, C., Rao, Y., Cheng, Y., Chen, C., Lu, J., Zhou, J.: Structure-preserving super resolution with gradient guidance. In: Proceedings of the IEEE Conference on Computer Vision and Pattern Recognition (CVPR) (2020)

14. Ma, C., Yang, C., Yang, X., Yang, M.: Learning a no-reference quality metric for single-image super-resolution. Comput. Vis. Image Underst. **158**, 1–16 (2017). https://doi.org/10.1016/j.cviu.2016.12.009
15. Martin, A., Soumith, C., Léon, B.: Wasserstein generative adversarial networks. In: Proceedings of the 34th International Conference on Machine Learning. Proceedings of Machine Learning Research, vol. 70, pp. 214–223. PMLR, 06–11 August 2017
16. Mescheder, L., Nowozin, S., Geiger, A.: Which training methods for GANs do actually converge? In: International Conference on Machine Learning (ICML) (2018)
17. Paszke, A., et al.: Pytorch: an imperative style, high-performance deep learning library (2019). https://doi.org/10.48550/ARXIV.1912.01703
18. Pham, C., Ducournau, A., Fablet, R., Rousseau, F.: Brain MRI super-resolution using deep 3d convolutional networks. In: 2017 IEEE 14th International Symposium on Biomedical Imaging (ISBI 2017), pp. 197–200 (2017). https://doi.org/10.1109/ISBI.2017.7950500
19. Sawyer, A.M., et al.: Creation of fully sampled MR data repository for compressed sensing of the knee. In: Proceedings of Society for MR Radiographers & Technologists (SMRT) 22nd Annual Meeting, Salt Lake City, UT, USA (2013)
20. Shi, W., et al.: Real-time single image and video super-resolution using an efficient sub-pixel convolutional neural network. In: Proceedings of the IEEE Conference on Computer Vision and Pattern Recognition (CVPR), June 2016
21. Sønderby, C.K., Caballero, J., Theis, L., Shi, W., Huszár, F.: Amortised MAP inference for image super-resolution. In: 5th International Conference on Learning Representations, ICLR 2017, OpenReview.net (2017). https://openreview.net/forum?id=S1RP6GLle
22. Sui, Y., Afacan, O., Gholipour, A., Warfield, S.K.: Learning a gradient guidance for spatially isotropic MRI super-resolution reconstruction. In: Martel, A.L., et al. (eds.) MICCAI 2020. LNCS, vol. 12262, pp. 136–146. Springer, Cham (2020). https://doi.org/10.1007/978-3-030-59713-9_14
23. Van Essen, D.C., Smith, S.M., Barch, D.M., Behrens, T.E.J., Yacoub, E., Ugurbil, K.: The WU-Minn human connectome project: an overview. Neuroimage **80**, 64–79 (2013)
24. Vandenhende, S., Brabandere, B.D., Neven, D., Gool, L.V.: A three-player GAN: generating hard samples to improve classification networks. CoRR abs/1903.03496 (2019)
25. Wang, X., et al.: ESRGAN: enhanced super-resolution generative adversarial networks. In: Leal-Taixé, L., Roth, S. (eds.) ECCV 2018. LNCS, vol. 11133, pp. 63–79. Springer, Cham (2019). https://doi.org/10.1007/978-3-030-11021-5_5
26. Wu, Q., et al.: An arbitrary scale super-resolution approach for 3d MR images via implicit neural representation. IEEE J. Biomed. Health Inform. **27**, 1–12 (2022)
27. Xu, K., Li, C., Zhu, J., Zhang, B.: Understanding and stabilizing GANs' training dynamics using control theory. In: Proceedings of the 37th International Conference on Machine Learning. Proceedings of Machine Learning Research, vol. 119, pp. 10566–10575. PMLR, 13–18 July 2020
28. Zhang, R., Isola, P., Efros, A., Shechtman, E., Wang, O.: The unreasonable effectiveness of deep features as a perceptual metric. In: CVPR (2018)
29. Zhang, Y., Li, K., Li, K., Fu, Y.: MR image super-resolution with squeeze and excitation reasoning attention network. In: 2021 IEEE/CVF Conference on Computer Vision and Pattern Recognition (CVPR), pp. 13420–13429 (2021). https://doi.org/10.1109/CVPR46437.2021.01322

30. Zhang, Y., Li, K., Li, K., Wang, L., Zhong, B., Fu, Y.: Image super-resolution using very deep residual channel attention networks. In: Ferrari, V., Hebert, M., Sminchisescu, C., Weiss, Y. (eds.) ECCV 2018. LNCS, vol. 11211, pp. 294–310. Springer, Cham (2018). https://doi.org/10.1007/978-3-030-01234-2_18

Cross-Attention for Improved Motion Correction in Brain PET

Zhuotong Cai[1,2,3], Tianyi Zeng[2(✉)], Eléonore V. Lieffrig[2], Jiazhen Zhang[3], Fuyao Chen[3], Takuya Toyonaga[2], Chenyu You[4], Jingmin Xin[1], Nanning Zheng[1], Yihuan Lu[6], James S. Duncan[2,3,4], and John A. Onofrey[2,3,5]

[1] Institute of Artificial Intelligence and Robotics, Xi'an Jiaotong University, Xi'an, China
zhuotongcai@gmail.com, {jxin,nnzheng}@mail.xjtu.edu.cn
[2] Department of Radiology and Biomedical Imaging, Yale University, New Haven, CT, USA
[3] Department of Biomedical Engineering, Yale University, New Haven, CT, USA
[4] Department of Electrical Engineering, Yale University, New Haven, CT, USA
[5] Department of Urology, Yale University, New Haven, CT, USA
{tianyi.zeng,jiazhen.zhang,eleonore.lieffrig,takuya.toyonaga}@yale.edu
[6] United Imaging Healthcare, Shanghai, China
yihuan.lu@united-imaging.com

Abstract. Head movement during long scan sessions degrades the quality of reconstruction in positron emission tomography (PET) and introduces artifacts, which limits clinical diagnosis and treatment. Recent deep learning-based motion correction work utilized raw PET list-mode data and hardware motion tracking (HMT) to learn head motion in a supervised manner. However, motion prediction results were not robust to testing subjects outside the training data domain. In this paper, we integrate a cross-attention mechanism into the supervised deep learning network to improve motion correction across test subjects. Specifically, cross-attention learns the spatial correspondence between the reference images and moving images to explicitly focus the model on the most correlative inherent information - the head region the motion correction. We validate our approach on brain PET data from two different scanners: HRRT without time of flight (ToF) and mCT with ToF. Compared with traditional and deep learning benchmarks, our network improved the performance of motion correction by 58% and 26% in translation and rotation, respectively, in multi-subject testing in HRRT studies. In mCT studies, our approach improved performance by 66% and 64% for translation and rotation, respectively. Our results demonstrate that cross-attention has the potential to improve the quality of brain PET image reconstruction without the dependence on HMT. All code will be released on GitHub: https://github.com/OnofreyLab/dl_hmc_attention_mlcn2023.

Supplementary Information The online version contains supplementary material available at https://doi.org/10.1007/978-3-031-44858-4_4.

Keywords: Motion Correction · Cross-attention · Brain · PET · Deep Learning

1 Introduction

Positron emission tomography (PET) is a widely used medical imaging technique that provides images of the metabolic activity of various tissues and organs within the human body [11,25]. It has shown remarkable clinical achievement in the study of physiological and pathological processes. However, the patient motion may lead to image quality degradation, decreased concentration in regions with high uptake, and incorrect outcome measures from kinetic analysis of dynamic datasets. In brain PET, swallowing, breathing, or involuntary movements can introduce significant artifacts like blurring or misalignment of structures that reduce the image resolution [10,12]. Thus, head motion correction is crucial in PET imaging to obtain better diagnosis and treatment planning for a wide range of medical conditions, such as Alzheimer's disease, Parkinson's disease, and brain tumors.

To overcome this challenge, researchers have developed various motion correction techniques specific to brain PET imaging that aim to improve the accuracy and reliability of the images. These techniques include physical head restraint, hardware motion tracking (HMT) [9], and data-driven methods that use the PET data itself [13,14,16,24]. HMT is not routinely used in the clinic, since the setup and calibration of the tracking device are complicated and the attachment of markers to each subject increases additional burden of the scan. Therefore, data-driven solutions are imperative and preferred, especially with the fast development of deep learning [2,4,5]. Revilla et al. [16] proposed an adaptive data-driven method using the center of tracer distribution (COD) to detect head motion. Zeng et al. [24] proposed a supervised deep learning framework to predict rigid transformation of head motion derived from the Polars Vicra, an external camera-based HMT. Lieffrig et al. [13] refined this approach by using a multi-task learning algorithm that combines cloud representation with the original training together where the network learns the point cloud representation embedding to improve the motion prediction performance. However, these methods showed limited performance when applied to the testing subjects dataset out of the training subjects dataset, especially on some subjects that exhibited large movements. The inherent noise of the 3D point cloud image (PCI) makes the above methods focus on noisy areas outside the head region, which leads to a poor generalization of data distribution and an inefficient learning of motion information.

In this paper, we propose a novel cross-attention mechanism to improve the performance of the model for multi-subject motion correction. Attention mechanisms have been shown to be effective for motion correction and image segmentation/registration in cardiac image analysis applications [1,3,7,20–22]. The cross-attention mechanism takes a pair of features as input and computes the correlations to build the spatial correspondence between the reference images and moving images. This explicitly enables the model to focus on the most important data information for motion correction – the head region. We evaluate motion correction results using both quantitative motion tracking comparison and qualitative reconstruction comparison with region of interest (ROI) evaluation to validate our method on two different PET scanners: an HRRT scanner without time of flight (ToF) and an mCT scanner with ToF. Results demonstrate the significant robustness and generalization improvement in motion correction of our method compared to traditional registration and deep learning benchmarks.

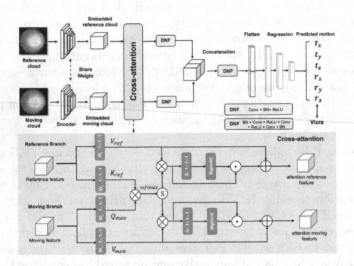

Fig. 1. Cross-attention network architecture. (Top) A shared encoder extracts the features from a pair of moving and reference clouds. Then, both obtained features are fed into the cross-attention module to learn the inherent correlation information. The concatenated attention features are fed into the regression layers to predict motion. (Bottom) Details of the cross-attention module.

2   Methodology

2.1   Dataset

In this work, we validate our approach using data from two different PET scanners with different imaging characteristics: 1) *HRRT Dataset* - 35 subjects with ^{18}F-FDG tracer were collected from a database acquired using a brain-dedicated High Resolution Research Tomograph (HRRT) scanner (Siemens, Erlangen, Germany) with resolution better than 3mm^3. The HRRT scanner does not include ToF information. 2) *mCT Dataset* - 6 subjects with ^{18}F-FPEB tracer were collected from a database acquired using a conventional ToF mCT PET-CT scanner (Siemens, Erlangen, Germany) with resolution of ~5mm^3 with 520 ps time resolution. All PET studies include raw list-mode data, Polaris Vicra (NDI, Canada) gold-standard motion tracking that records motion at a rate up to 30 Hz, and T1 magnetic resonance imaging (MRI). To reduce confounds caused by PET tracer uptake distribution during early time frames, we consider only imaging data from the 30 min window 60–90 min post-injection.

2.2 Proposed Network

We propose an end-to-end network to predict rigid head motion in PET imaging. This network contains three parts (Fig. 1). The feature extractor adopts a shared-weight convolutional encoder to capture the regional feature of the reference image $I_{ref} \in \mathbb{R}^3$ and moving image $I_{mov} \in \mathbb{R}^3$. Then, the proposed cross-attention block captures both local features and their global correspondence between the reference and moving images. Finally, the processed feature information is fed into a fully connected multilayer perceptron (MLP) block to infer the six rigid transformation motion parameters $\theta = [t_x, t_y, t_z, r_x, r_y, r_z]$ (three translation t_d and rotation r_d parameters in each axis d).

Cross-Attention: We aim to improve motion correction performance by having the model learn to emphasize the head region of interest. To explicitly discriminate the head region, we propose the cross-attention module (Fig. 1) to establish spatial correspondences between features from the reference image and the moving image. The two inputs to the cross-attention module are denoted as $f_{ref} \in R^{C \times H \times W \times L}$ and $f_{mov} \in R^{C \times H \times W \times L}$ for the feature maps of the reference and moving images, respectively, where H, W, L indicate the feature map size and C indicates the feature channel. We first divide the input reference feature f_{ref} into reference key Q_{ref} and value V_{ref} and the input moving feature f_{mov} into moving query Q_{mov} and value V_{mov}, which are defined as:

$$Q_{ref} = W_a f_{ref}, \quad V_{ref} = W_b f_{ref}$$
$$Q_{mov} = W_a f_{mov}, \quad V_{mov} = W_b f_{mov}.$$

We reshape Q_{mov} and Q_{ref} to the dimension of $C \times (HWL)$ and calculate the attention matrix using the following equation:

$$A_{mr} = \text{Softmax}(Q_{mov}^T Q_{ref}) \in R^{(HWL) \times (HWL)}$$

where A_{mr} reflects the similarity between each row of Q_{mov}^T and each column of Q_{ref}. Once the attention map A_{mr} is calculated, the updated attentions on the reference feature and moving feature are as follows:

$$A_{ref} = A_{mr} \cdot V_{ref}, \quad A_{mov} = A_{mr}^T \cdot V_{mov}.$$

To better determine which parts of the moving image are most relevant for motion with the reference image and how much influence those parts should have on the final motion prediction, we utilize the self-gate mechanism to weight the information from different inputs. It helps the model to selectively combine information from the moving image with the reference image. The gate is formulated as follows:

$$G_{ref} = \sigma(A_{ref}) \in [0,1]^{HWL}, \quad G_{mov} = \sigma(A_{mov}) \in [0,1]^{HWL}$$

where σ is the logistic sigmoid activation function. The gate G determines how much information from the reference image will be preserved and can be learned automatically. The final attention feature representation for the moving feature and reference feature can be obtained by:

$$F_{ref} = G_{ref}A_{ref} + V_{ref}, \quad F_{mov} = G_{mov}A_{mov} + V_{mov}$$

Thus, through the cross-attention mechanism, attention will be gradually focused on much more relevant information, while redundant features will be gradually suppressed, greatly improving the robustness and accuracy of the motion correction.

Network Architecture: Instead of using ResNet for the feature extractor in [24], we adopt a shallow convolutional encoder network [13] to extract the features from the moving and reference images. This shallow encoder uses 3 convolution layers with kernel size 5^3 and one intermediate convolution layer with kernel size of 1 and number of features 32, 64, 128 with batch normalization and ReLU layers. Additionally, we integrate a deep normalization and fusion module (DNF) [7] with a sequence of batch normalization, ReLU, and convolution layers before and after the concatenation to enhance the feature representation of the information aggregation. Our network architecture also does not utilize a time conditioning block [24]. The network has 9.32M trainable parameters.

2.3   Implementation Detail

For the training procedure, we randomly select samples where the time points of reference and moving satisfy $\{t_{ref}, t_{mov} \in (0, 1800)|t_{ref} < t_{mov}\}$. For training, Adam optimizer to train the whole model with the learning rate and weight decay of 5e-4 and 0.98, separately. For the inference procedure, we fix $t_{ref} = 0$ at the beginning and predict motion for all following $t_{mov} > 0$. The model was trained with PyTorch (v1.7.1) and MONAI (v0.9) on a workstation with Intel Xeon Gold 5218 processors and 256 GB RAM and NVIDIA Quadro RTX 8000 GPU (48 GB RAM). All code will be made available on GitHub upon acceptance.

3   Experimental Results

We validate our approach to brain PET head motion correction through quantitative and qualitative evaluation, ablation studies, and show the interpretability of our results. For the HRRT study, we utilize 30 subjects as the training set and 5 subjects for the testing set. For the mCT study, we use 5 subjects for training and 1 subject for testing. From the training set images, 10% of the data is reserved as a validation set to measure model convergence. Following [24], one-second PCI representations of PET data are adopted as the inputs for the model. Due to GPU memory limitation, HRRT PCIs were preprocessed and downsampled to size of $32 \times 32 \times 32$ voxels ($9.76 \times 9.76 \times 7.96\,\text{mm}^3$) and mCT PCIs were resized to $96 \times 96 \times 64$ voxels ($3.18 \times 3.18 \times 3.45\,\text{mm}^3$).

For the HRRT dataset, we compared our approach to head motion correction with several motion correction methods: (i) BioImage Suite [15] (BIS), a traditional intensity-based rigid registration procedure that minimizes the sum-of-squared differences between the PCIs; (ii) DL-HMC [24] supervised learning for head motion correction without attention; (iii) mtDL-HMC [13] a multi-task learning version of DL-HMC; and (iv) we verify the utility of the cross-attention module by replacing it with the original self-attention module [19]. For the mCT dataset, we compared our approach to DL-HMC [24]. All models were trained using mini-batches of size 256 on HHRT and 8 on mCT for 10k epochs until the model converged on the validation set.

Motion Correction Evaluation: We calculated Mean Square Error (MSE) between the Vicra gold-standard motion and the predicted transformation in translation and rotation separately, and use Student's t-test to assess significance ($p < 0.05$). Table 1 shows quantitative results from the validation and test sets for the HRRT and mCT datasets. We observe that our proposed method significantly outperforms the other methods in both datasets. Compared to the benchmark DL-HMC, our approach demonstrates a 58% and 26% improvement in translation and rotation performance, respectively, for the HRRT dataset and a 66% and 64% improvement in the mCT dataset. The results between self-attention and our method indicate that cross-attention can better learn the shared inherent feature information through the attention matrix computed from reference and moving features. The failure of BIS highlights the challenging nature of this motion correction task using the low-resolution PCI data representation.

Figure 2 presents motion prediction results from the HRRT and mCT test sets. Compared with DL-HMC, we observe that our proposed method can more accurately and effectively capture the motion both in translation and rotation throughout the scan period, especially for a subject with large movements (Fig. 2 first row). In this case, DL-HMC performance degrades when there is a sudden movement during rotation and translation. Although DL-HMC captured the movement trends, it failed to quantify the magnitude of the movement. The attention model more closely approximates the Vicra gold-standard, demonstrating that our method can better capture large motion movements through the attention mechanism. mCT motion prediction results (Fig. 2 third row) demonstrate the superior performance of our method using a different scanner with ToF.

Table 1. Quantitative motion prediction results. Motion prediction translation and rotation transformation error MSE (mean±SD) compared to Vicra gold-standard motion tracking on the HRRT and mCT datasets.

	Method	Validation Set		Test Set	
		Trans. (mm)	Rot. (deg)	Trans. (mm)	Rot. (deg)
HRRT	BIS [15]	-	-	6.68 ± 11.70	2.82 ± 3.47
	DL-HMC [24]	1.81 ± 8.52	1.03 ± 3.69	4.08 ± 9.32	1.76 ± 3.04
	mtDL-HMC [13]	-	-	2.31 ± 5.62	1.57 ± 2.43
	Self-attention [19]	0.37 ± 2.93	0.30 ± 0.78	1.97 ± 5.11	1.47 ± 2.49
	Proposed	**0.33 ± 0.96**	**0.25 ± 0.66**	**1.72 ± 4.68**	**1.30 ± 2.17**
mCT	DL-HMC [24]	1.40 ± 2.97	0.67 ± 1.37	2.90 ± 1.58	0.42 ± 0.35
	Proposed	**0.49 ± 1.54**	**0.21 ± 0.39**	**0.98 ± 0.58**	**0.15 ±0.16**

Ablation Studies: Experiments on the HRRT dataset verified the effectiveness of the self-gate mechanism and DNF modules in our method (Table 2). Each part positively contributes to our model's performance, especially the DNF. Removing the DNF degrades performance greatly in both translation and rotation.

PET Reconstruction Evaluation and Model Interpretability: Motion compensation OSEM List-mode Algorithm for Resolution-Recovery Reconstruction (MOLAR) [6] was used to reconstruct the PET images from predicted motion correction information during the whole sequence. We compared reconstruction using no motion correction (NMC), DL-HMC, our proposed method, and the Vicra gold-standard (Fig. 3(a)). Qualitatively, the attention-based model demonstrates improvement in definition of important brain anatomical struc-

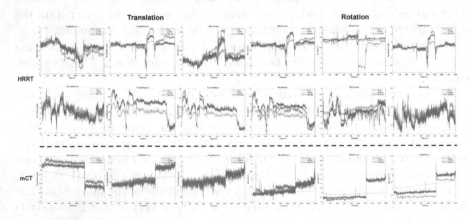

Fig. 2. Motion prediction results. Columns show rigid transformation parameters (from left to right: translation in x, y, z directions and rotation about the x, y, z axes) from gold-standard Vicra (Orange), DL-HMC (Yellow) and our proposed method (Blue) on two different PET scanners: two test subjects from HRRT (without ToF) and one subject from mCT (with ToF).

tures. For instance, the attention-based method allows better delineation of brain gyrus and sulcus as compared to NMC and baseline DL-HMC model results, and are indiscernible from Vicra results in this example subject. We also observe substantial improvement in visualizing the separation and connection of the two brain hemispheres by highlighting the longitudinal fissure (grey matter) and corpus callosum (white matter). Figure 3(a) illustrates a PET image of a subject with epilepsy. While it is difficult to differentiate on both NMC and baseline DL-HMC model results, the attention-based model reveals evident glucose hypometabolism in the right frontal cortex that likely corresponds to origin of seizure activity and agrees with the finding in the Vicra-based gold-standard results. Therefore, we further demonstrate the promising advantage of the attention-based method in the identification of potential pathology on PET images through motion correction.

To quantitatively assess the reconstructed PET images, each subject's co-registered MRI was segmented into 12 grey matter brain regions of interest (ROIs) using FreeSurfer [8] and analyzed by calculating the difference in ROI Standard Uptake Value (SUV) compared to the Vicra gold-standard result. We averaged the SUV results across all 5 testing subjects in the HRRT dataset (Fig. 3(b)). For the FDG tracer used in this study, grey matter ROIs should have higher activity. Our proposed attention-based method outperformed DL-HMC over all ROIs, demonstrating smaller errors from the Vicra gold-standard and less variation. Overall, our method yielded SUV results similar to the Vicra gold-standard, showing promise for clinical diagnosis of brain disease.

To better understand why the attention module can improve the performance of the motion correction, we applied Grad-CAM [17] to the model to highlight the region of interest in model and compared to saliency maps of DL-HMC. Grad-CAM results from a subject in the test set illustrate the effectiveness of our approach to focus on the head region (Fig. 4) at three time points of moving images at the beginning, middle and end of the scane, t=143, 804 and 1,716 s, respectively. On the other hand, saliency maps highlight regions outside of the head using DL-HMC , which indicates that the model failed to focus on the relevant anatomical information in the PCI.

4   Discussion and Conclusion

In this study, we demonstrate that cross-attention significantly improves motion prediction performance in brain PET. Grad-CAM results help to explain why our proposed solution achieves more robust motion correction performance on the test set, especially in subjects with large movements. Our results also demonstrate the successful application of our approach to two different PET scanners with distinct imaging characteristics, which differs from previous work [13,24] that performed motion correction on a single scanner. While the complexity of the imaging data and supervised training process limited the application of our motion correction method to a small number of test subjects, a single test subject for the mCT data, our results are highly encouraging. Initial results indicate that improved data-driven motion correction performance can be realized

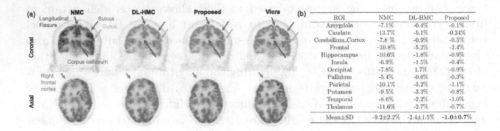

Fig. 3. PET image reconstruction results. (a) MOLAR reconstructed images using no motion correction (NMC), DL-HMC, our proposed cross-attention method and Vicra gold-standard motion tracking for a single subject with epilepsy. (b) Relative change in SUV in 12 brain ROIs comparing NMC, DL-HMC, and our proposed compared to Vicra gold-standard across 5 subjects.

by PET scanners employing ToF information (e.g. mCT), which has the potential to generate high quality PCI. With the help of parallel computing with GPU, we could also achieve ultra-fast reconstruction for short frames like PCI [18,23]. Further study including big cohort study which is necessary to reach definitive conclusions about the benefits of ToF for brain PET deep learning head motion correction, and enhancing PCI quality via PET reconstruction techniques.

S   Supplementary Material

S.1   Ablation Study Results

Table 2. Ablation study on HRRT. Motion prediction transformation error MSE (mean±SD) compared to Vicra gold-standard motion tracking.

Method	Validation Set Translation (mm)	Rotation (deg)	Test Set Translation (mm)	Rotation (deg)
Proposed	**0.33 ± 0.96**	**0.25 ± 0.66**	**1.72 ± 4.68**	**1.30 ± 2.17**
w/o gate	0.37 ± 2.94	0.29 ± 0.78	2.00 ± 4.90	1.49 ± 2.45
w/o DNF	0.73 ± 3.31	0.47 ± 1.78	2.52 ± 5.95	1.78 ± 2.97

S.2   Model Interpretability

Acknowledgements. This work was supported by the National Key Research and Development Program of China 2017YFA0700800 and the National Institute of Health (NIH) R21 EB028954.

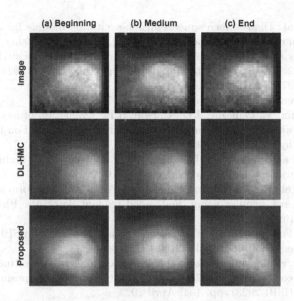

Fig. 4. Grad-CAM saliency map visualization. Sagittal view from three different time frames from the HRRT test set at the (a) beginning (t=143 s), (b) middle (t=804 s), and (c) end (t=1,716 s) of the 30 min (1,800 s) PET acquisition. Our proposed method more accurately localizes the head anatomy compared to the DL-HMC method does not utilize an attention mechanism.

References

1. Ahn, S.S., et al.: Co-attention spatial transformer network for unsupervised motion tracking and cardiac strain analysis in 3D echocardiography. Med. Image Anal. **84**, 102711 (2023)
2. Cai, Z., Xin, J., Liu, S., Wu, J., Zheng, N.: Architecture and factor design of fully convolutional neural networks for retinal vessel segmentation. In: 2018 Chinese Automation Congress (CAC), pp. 3076–3080. IEEE (2018)
3. Cai, Z., Xin, J., Shi, P., Wu, J., Zheng, N.: DSTUnet: Unet with efficient dense Swin transformer pathway for medical image segmentation. In: 2022 IEEE 19th International Symposium on Biomedical Imaging (ISBI), pp. 1–5. IEEE (2022)
4. Cai, Z., Xin, J., Shi, P., Zhou, S., Wu, J., Zheng, N.: Meta pixel loss correction for medical image segmentation with noisy labels. In: Zamzmi, G., Antani, S., Bagci, U., Linguraru, M.G., Rajaraman, S., Xue, Z. (eds.) Medical Image Learning with Limited and Noisy Data. MILLanD 2022. LNCS, vol. 13559, pp. 32–41. Springer, Cham (2022). https://doi.org/10.1007/978-3-031-16760-7_4
5. Cai, Z., Xin, J., Wu, J., Liu, S., Zuo, W., Zheng, N.: Triple multi-scale adversarial learning with self-attention and quality loss for unpaired fundus fluorescein angiography synthesis. In: 2020 42nd Annual International Conference of the IEEE Engineering in Medicine & Biology Society (EMBC), pp. 1592–1595. IEEE (2020)
6. Carson, R.E., Barker, W.C., Liow, J.S., Johnson, C.A.: Design of a motion-compensation OSEM list-mode algorithm for resolution-recovery reconstruction for the HRRT. In: 2003 IEEE Nuclear Science Symposium. Conference Record (IEEE CAT. No. 03CH37515), vol. 5, pp. 3281–3285. IEEE (2003)

7. Chen, X., et al.: Dual-branch squeeze-fusion-excitation module for cross-modality registration of cardiac Spect and CT. In: Wang, L., Dou, ., Fletcher, P.T., Speidel, S., Li, S. (eds.) Medical Image Computing and Computer Assisted Intervention. MICCAI 2022. LNCS, vol. 13436 pp. 46–55. Springer, Cham (2022). https://doi.org/10.1007/978-3-031-16446-0_5

8. Fischl, B.: FreeSurfer. Neuroimage **62**(2), 774–781 (2012)

9. Jin, X., Mulnix, T., Gallezot, J.D., Carson, R.E.: Evaluation of motion correction methods in human brain pet imaging-a simulation study based on human motion data. Med. Phys. **40**(10), 102503 (2013)

10. Kuang, Z., et al.: Progress of SIAT bPET: an MRI compatible brain PET scanner with high spatial resolution and high sensitivity (2022)

11. Kuang, Z., et al.: Design and performance of SIAT aPET: a uniform high-resolution small animal pet scanner using dual-ended readout detectors. Phys. Med. Biol. **65**(23), 235013 (2020)

12. Kyme, A.Z., Fulton, R.R.: Motion estimation and correction in SPECT PET and CT. Phys. Med. Biol. **66**(18), 18TR02 (2021)

13. Lieffrig, E.V., et al.: Multi-task deep learning and uncertainty estimation for PET head motion correction. In: 2023 IEEE 20th International Symposium on Biomedical Imaging (ISBI 2023), pp. 1–4, April 2023

14. Lu, Y., et al.: Data-Driven motion detection and event-by-event correction for brain PET: comparison with Vicra. J. Nucl. Med. **61**(9), 1397–1403 (2020)

15. Papademetris, X., Jackowski, M.P., Rajeevan, N., DiStasio, M., Okuda, H., Constable, R.T., Staib, L.H.: Bioimage suite: an integrated medical image analysis suite: an update. Insight J. **2006**, 209 (2006)

16. Revilla, E.M., et al.: Adaptive data-driven motion detection and optimized correction for brain pet. Neuroimage **252**, 119031 (2022)

17. Selvaraju, R.R., Cogswell, M., Das, A., Vedantam, R., Parikh, D., Batra, D.: Grad-CAM: visual explanations from deep networks via gradient-based localization. In: Proceedings of the IEEE International Conference on Computer Vision, pp. 618–626 (2017)

18. Spangler-Bickell, M.G., Deller, T.W., Bettinardi, V., Jansen, F.: Ultra-fast listmode reconstruction of short pet frames and example applications. J. Nucl. Med. **62**(2), 287–292 (2021)

19. Vaswani, A., et al.: Attention is all you need. In: Advances in Neural Information Processing Systems, vol. 30 (2017)

20. You, C., Dai, W., Min, Y., Staib, L., Duncan, J.S.: Implicit anatomical rendering for medical image segmentation with stochastic experts. arXiv preprint arXiv:2304.03209 (2023)

21. You, C., et al.: Incremental learning meets transfer learning: application to multi-site prostate MRI segmentation. In: Albarqouni, S., et al. (eds.) Distributed, Collaborative, and Federated Learning, and Affordable AI and Healthcare for Resource Diverse Global Health. DeCaF FAIR 2022 2022. LNCS, vol. 13573, pp. 3–16. Springer, Cham (2022). https://doi.org/10.1007/978-3-031-18523-6_1

22. You, C., et al.: Class-aware adversarial transformers for medical image segmentation. In: Advances in Neural Information Processing Systems (2022)

23. Zeng, T., et al.: A GPU-accelerated fully 3d Osem image reconstruction for a high-resolution small animal pet scanner using dual-ended readout detectors. Phys. Med. Biol. **65**(24), 245007 (2020)

24. Zeng, T., et al.: Supervised deep learning for head motion correction in pet. In: Wang, L., Dou, Q., Fletcher, P.T., Speidel, S., Li, S. (eds.) Medical Image Comput-

ing and Computer Assisted Intervention-MICCAI 2022: 25th International Conference, Singapore, September 18–22, 2022, Proceedings, Part IV, pp. 194–203. Springer, Cham (2022). https://doi.org/10.1007/978-3-031-16440-8_19

25. Zeng, T., et al.: Design and system evaluation of a dual-panel portable pet (DP-PET). EJNMMI Phys. **8**(1), 1–16 (2021)

VesselShot: Few-shot Learning
for Cerebral Blood Vessel Segmentation

Mumu Aktar[1]([✉])[iD], Hassan Rivaz[2][iD], Marta Kersten-Oertel[1][iD],
and Yiming Xiao[1][iD]

[1] Department of Computer Science and Software Engineering, Concordia University,
Montreal, Canada
m_ktar@encs.concordia.ca
[2] Department of Electrical and Computer Engineering, Concordia University,
Montreal, Canada

Abstract. Angiography is widely used to detect, diagnose, and treat cerebrovascular diseases. While numerous techniques have been proposed to segment the vascular network from different imaging modalities, deep learning (DL) has emerged as a promising approach. However, existing DL methods often depend on proprietary datasets and extensive manual annotation. Moreover, the availability of pre-trained networks specifically for medical domains and 3D volumes is limited. To overcome these challenges, we propose a few-shot learning approach called "VesselShot" for cerebrovascular segmentation. VesselShot leverages knowledge from a few annotated support images and mitigates the scarcity of labeled data and the need for extensive annotation in cerebral blood vessel segmentation. We evaluated the performance of VesselShot using the publicly available TubeTK dataset for the segmentation task, achieving a mean Dice coefficient (DC) of 0.62 ± 0.03.

Keywords: Few-shot learning · Deep learning · 3D volumes · Cerebrovascular segmentation

1 Introduction

Cerebral blood vessel segmentation plays a vital role in various applications, such as diagnosing cerebrovascular diseases (e.g., stroke), planning surgical interventions for conditions, such as aneurysms and arteriovenous malformations, studying brain functions, and assessing the impact of new treatments for cerebrovascular disorders. However, publicly labeled data in this domain is limited, hindering the progress of deep learning-based research on cerebral blood vessel segmentation. Since 2017, numerous deep learning-based methods have been proposed for cerebral blood vessel segmentation. However, most of the previous work has been performed on data from private sources as there is very little publicly available labeled data [1]. Traditionally, deep learning (DL) models for semantic segmentation require a large amount of training data with manual annotation, which is time-consuming and labor-intensive, particularly in 3D, where many slices must be inspected. The need for more annotated data negatively impacts DL models' training and generalization capabilities.

A. Abdulkadir et al. (Eds.): MLCN 2023, LNCS 14312, pp. 46–55, 2023.
https://doi.org/10.1007/978-3-031-44858-4_5

To address this challenge, few-shot learning emerges as a promising alternative that reduces the need for extensive manual annotation. For semantic image segmentation, few-shot learning aims to enable DL models to learn underlying visual patterns and semantics from a limited set of labeled examples. This also allows them to generalize effectively to unseen object categories during the segmentation process. To date, few-shot segmentation has been explored in several medical imaging contexts. In the study of Roy *et al.* [2], the authors proposed a two-armed few-shot architecture to extract support and query images with squeeze-and-excitation modules for the segmentation of abdominal organs (each organ type is considered as a separate class) using 3D volumetric scans, obtaining an average Dice coefficient (DC) of 48.5% [2]. Similarly, Tang *et al.* [3] used a few-shot framework to refine the segmentation masks with a recurrent module and achieved a mean DC of 81.91%. To eliminate expert annotation for training medical image segmentation algorithms, Ouyang *et al.* [4] employed a super-pixel-based self-supervised segmentation approach with few-shot learning. Preserving the local information to alleviate the foreground vs. background imbalance issue with an adaptive local prototype pooling, their study achieved a maximum DC of 78.84%. Semi-supervised segmentation was also incorporated in a few-shot paradigm by considering a generative adversarial network (GAN) [5]. This method had comparable performance to fully supervised approaches in multi-modal 3D medical image segmentation. Few-shot learning techniques have also shown efficacy in cardiac image sequence segmentation tasks following a multi-level semantic adaptation, with a DC of 92.43% [6]. Lastly, Xu *et al.* [7] proposed a few-shot learning method with a multi-scale class prototype and attention module for 2D retinal blood vessel segmentation.

In this paper, we aimed to develop a few-shot learning approach, called "VesselShot", for segmenting cerebral blood vessels. Building upon the PANet few-shot segmentation method introduced by Wang *et al.* [8] for natural image segmentation based on metric learning, VesselShot leverages DL models' ability to learn a consistent embedding space that minimizes the distance between support and query prototypes (see Sect. 2.2). To the best of our knowledge, our method is the first attempt to employ few-shot learning for 3D segmentation of brain vascular images. The proposed VesselShot technique aims to overcome the limitations of the existing deep-learning models for cerebral blood vessel segmentation and explore the potential of few-shot learning in this domain.

2 Methodology

2.1 Dataset and Pre-processing

We used the publicly available TubeTK dataset[1]. Among the 100 magnetic resonance angiography (MRA) of healthy subjects, a subset of 42 have manual segmentation of the intracranial vasculature. The original dimension of the images is $448 \times 448 \times 128$ voxels at a resolution of $0.5 \times 0.5 \times 0.8$ (mm^3). The images

---

[1] https://public.kitware.com/Wiki/TubeTK/Data.

were pre-processed as follows. First, all images were down-sampled to a resolution of $1 \times 1 \times 1$ (mm^3), resulting in a dimension of $230 \times 230 \times 102$ voxels. To allow spatial consistency, all the brains were registered to one subject's image as a template with affine transformations. Fifteen patches that contain blood vessels were randomly extracted from each brain using the technique introduced in the study of Wang et al. [9], with a size of $64 \times 64 \times 16$ voxels to fit the GPU memory.

2.2 Problem Definition

To segment cerebral blood vessels with a small amount of annotated training data, we built upon the few-shot segmentation method proposed by Wang et al. [8]. In general, few-shot learning involves training and testing episodes with support and query sets, following a "C-way K-shot" paradigm. The support set comprises labeled examples that a DL model can use to learn about target classes, while the query set contains unseen test cases to be classified during inference. In C-way K-shot segmentation, we obtain K {image, mask} pairs per semantic class in the support set, with a total of C classes. The training episodes consist of $S_{i,k}$, $M_{i,k}$ and $Q_{i,k}$, denoting support, mask, and query sets, respectively with $i = 1, 2, ...c$ for c classes and $k = 1, 2, ...s$ for s samples/shots. Both the support and query sets share knowledge extracted by the DL model to perform the final segmentation. In our experiments, following the problem framing of Roy et al. [2], who categorized classes with the designated segmentation tasks, we primarily focused on building our algorithm based on one class (i.e., blood vessel segmentation) or a 1-way K-shot approach. Furthermore, to account for individual vascular differences between subjects, we also considered the problem framing from the work of Xu et al. [7], who treated each subject as a separate class with its image patches as members of the class for retinal vessel segmentation. In this case, the segmentation was extended to a C-way K-shot setting.

2.3 Model Design

To perform support-to-query segmentation, we built robust prototypes from the target class of the support set. We used the nn-UNet [10] architecture as a backbone network to extract deep features from support and query images. Upsampling along with masked average pooling [8] was performed to obtain the final segmentation. Figure 1 shows an example of a 1-way 3-shot learning paradigm for query mask generation.

To generate a prototype from a class, the feature map, F_I of an image, I was extracted from the support set, S and by masked average pooling (Eq. 1), the obtained feature maps were compared across different class indices of I, similar to the prototype extraction approach used in PANet [8].

$$prototype_{\text{class}} = \frac{1}{N_s} \sum_{i=1}^{N_s} \frac{\sum_{x,y,z} F_{I,\text{class}}^{(x,y,z)} \mathbb{1}[Mask_{I,\text{class}}^{(x,y,z)} == \text{class}]}{\sum_{x,y,z} \mathbb{1}[Mask_{I,\text{class}}^{(x,y,z)} == \text{class}]} \tag{1}$$

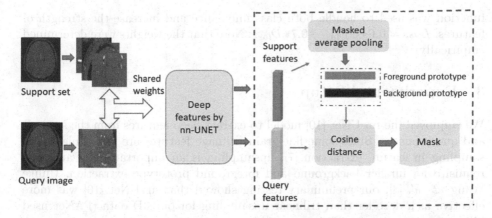

Fig. 1. VesselShot 1-way 3-shot learning: 1 brain with 3 sample patches is considered for the support set. Knowledge is shared between support and query set by extracting deep features using nn-UNet [10] which are further embedded into foreground and background prototypes using masked average pooling [8]. Cosine similarity is used between support and query prototypes to obtain the segmented query mask. (Color figure online)

where N_s is the total number of support images, and $\mathbb{1}[\cdot]$ takes a value of 1 when the condition $[Mask_{I,class}^{(x,y,z)} == class]$ is true and 0 otherwise. The background prototype was built following the same equation with the constraint of $[Mask_{I,class} \neq class]$, which means the feature map values do not belong to the corresponding class index. For the evaluation, all brain class indices were assigned a value of 1 for foreground blood vessels (our main target) and 0 for the background. After prototype extraction, the feature map of the query image was compared with the support prototypes using cosine similarity. Each voxel at the spatial location (x, y, z) was classified based on the maximum similarity between the query feature map and the support prototypes. Finally, a segmentation mask was predicted for the query image based on the maximum probability values obtained by Softmax that was applied to the distance map. Note that our 3D segmentation was performed similarly to the work of Roy *et al.* [2], but we experimented with the scenarios of 1-way K-shot and C-way K-shot as mentioned in Sect. 2.2. During inference time, we extracted 54 non-overlapping patches from a test MRA, where each was a query image at a given time and was paired with the support set to obtain the final segmentation. For our experiments, the support sets were created from the training data.

For training, a combination of both cross-entropy loss, CE_{loss} and Dice loss, D_{loss} was used. Our approach emphasized the Dice loss since blood vessels occupy a minimal area considering the brain space, which can significantly affect learning with a high-class imbalance. The Dice loss only focuses on the agreement of an image's predicted segmentation and ground truth (GT) label. However, it is not ideal to overlook the background entirely, as this can affect the robustness of significant features of the network [11]. Therefore, the following hybrid loss function was used to handle both class imbalance and increase the strength of features: $Loss = 0.6 * CE_{loss} + 0.7 * D_{loss}$. Note that the weights were determined empirically.

3   Experimental Setup

We employed the nn-UNet [10] model to extract deep features from the support and query images. Since some fine-grained image features are lost while downsampling in feature extraction [7], upsampling is an important step and prerequisite for further background and foreground prototype extraction. Unlike Wang *et al.* [8], our preliminary testing showed that nn-UNet [10] was more effective as a decoder than trilinear upsampling for our 3D data (PANet used bilinear upsampling for 2D images [8]).

To perform training in our few-shot blood vessel segmentation, a maximum iteration of 20,000 was used while monitoring the best DC value for early stopping. A learning rate (lr) scheduler with an SGD optimizer was used at an initial lr=0.01 and momentum of=0.99. For data augmentation, random flipping, random Gaussian blurring, noise addition, and contrast changes were applied during training. To evaluate the performance of our method, different metrics, including DC, precision, sensitivity, and Intersection over Union (IoU) were computed. We performed multiple experiments to prove the efficacy of the proposed fewshot segmentation method. These experiments encompassed various settings, namely 1-way 1-shot, 1-way 4-shot, 1-way 5-shot, and 3-way 5-shot learning. In addition, we also used a fully supervised UNet with 4 layers of hierarchies as a baseline to assess the performance of the proposed method using the same patch size and data augmentation techniques. It was trained using the Dice loss with an Adam optimizer ($lr = 0.0001$) and a CosineAnnealingLR scheduler ($T_{max} = 5, eta_{min} = 0.000001$). To compare the different few-shot learning settings, as well as the UNet baseline, we divided the data into three subsets: a training set (78%, 33 cases), a validation set (7%, 3 cases), and a test set (15%, 6 cases). The best setting was determined from their performance based on the test set. Subsequently, we used the best setting to conduct a full 4-fold crossvalidation. This way, we could obtain segmentation results for all the subjects to offer a more comprehensive evaluation.

4   Results

Table 1 presents the performance for various few-shot segmentation settings. It is important to note that the reported performance in the table was obtained from patch-based evaluations, where the averages of all classes in the test set were considered. The highest average DC of 0.67 was obtained with 1-way 1-shot learning. In terms of the mean values, 1-way 4-shot and 1-way 5-shot give similar results in all the evaluation criteria. However, the performance of the 3-way 5-shot method was notably inferior. This discrepancy may be attributed to the inclusion of three classes represented by distinct brains, characterized by significant similarities. Excess prototype generation in this approach likely contributed to overfitting, resulting in the observed decline in performance. The results of the 1-way 1-shot proved that the few-shot paradigm could offer sufficient segmentation performance with even a single sample from a single class, which makes faster convergence and mitigates the issue of a small annotated dataset.

It is essential to note that while the single-split result indicated a slight advantage for the 1-way 1-shot model in the case of DC, the more comprehensive evaluation through cross-validation provided a more precise and more reliable picture of the model's performance. Therefore, considering all performance metrics, the 1-way 5-shot setting emerged as the top-performing setting among the options tested in this study. Based on the 1-way 5-shot setting, a full 4-fold cross-validation was performed. The metrics of DC, sensitivity, precision, and IoU were obtained as 0.62 ± 0.03, 0.53 ± 0.02, 0.72 ± 0.02, and 0.43 ± 0.02, respectively. For qualitative evaluation, segmentation maps of four random patches are shown in Fig. 2 for the 1-way 5-shot setting. Furthermore, by recombining segmented image patches from the same brain spatially, we also demonstrate a case in Fig. 3 for the same setting. Finally, the fully supervised UNet performed poorly with the limited annotated data and achieved a DC of 0.27 ± 0.27. UNet typically requires a larger well-labeled dataset to achieve reasonable performance, as demonstrated by the study of Livne et $al.$ [12].

Table 1. Performance metrics of VesselShot for different settings with the UNet as a baseline, including DC, Sensitivity, Precision, and IoU.

Methods	DC (SD)	Sensitivity(SD)	Precision(SD)	IoU(SD)
1-way 1-shot	**0.67(0.02)**	0.50 (0.03)	0.68 (0.02)	0.40 (0.02)
1-way 4-shot	0.66 (0.02)	0.54 (0.03)	**0.68 (0.02)**	0.41 (0.02)
1-way 5-shot	0.66 (0.02)	0.58 (0.02)	**0.71 (0.03)**	**0.45 (0.02)**
3-way 5-shot	0.52 (0.04)	0.39 (0.04)	0.57 (0.08)	0.23 (0.03)
UNet	0.27 (0.27)	0.47 (0.09)	0.35 (0.06)	0.15 (0.04)

Fig. 2. Segmentation maps of four samples, where red represents the original cerebral blood vessels, blue shows the prediction, and purple represents the overlap (Color figure online)

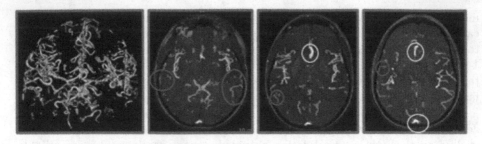

Fig. 3. *From left to right:* 3D segmentation result with the overlap of GT and predicted labels (yellow = overlap, green = GT and red = prediction). 2D Maximum intensity projections (MIPs) of 10 slices in Slice 15–25, 25–35, and 35–45 from a total of 102 brain slices, with the overlay of the original MRA and segmentation (in red). The blue circles show the wrong prediction of large vessels and the yellow circles indicate missed blood vessels. (Color figure online)

5   Discussion

This paper proposed a novel few-shot learning approach for 3D cerebral blood vessel segmentation. The method achieves the segmentation by building robust prototypes with masked average pooling based on embedded features that are extracted from an nn-UNet [10]. Inspired by the PANet [8], we adopted foreground and background prototypes and used them to compare the query feature map with the support set's prototypes for blood vessel segmentation. To further enhance generalizability, Wang et al. [8] performed prototype alignment regularization (PAR) of the predicted query mask with the support mask through additional information extraction. We have also experimented with this technique for our application, but unfortunately, it did not lead to performance gains during evaluation and resulted in slower convergence in model training.

In our approach, we considered two scenarios: (1) a single class (blood vessel segmentation) in a 1-way K-shot setting and (2) a C-way K-shot setting with each brain as a separate class. While the first case aligns with the approach of Roy et al. [2], the latter resembles the problem framing of Xu et al. [7]. In the 1-way K-shot setting, the best results came from the 1-way 5-shot setting, which was superior to treating different brains as their own classes. We hypothesize that this was due to the high structural similarity between MRAs after spatial normalization, which emphasizes the primary vasculature networks. This is in contrast to the results of Xu et al. [7]. Since the core task involves only two classes, blood vessels, and the background, the C-way K-shot paradigm may lead to overfitting and compromise performance. In the future, we will continue to explore different framings of the C-way K-shot setup for improved accuracy. For example, treating 3D image patches from consistent spatial locations in a stereotactic space as distinct classes to allow enhanced feature encoding.

We found that misclassifications predominantly occurred near the brain surface, where surface veins and the dura (both with bright signals) reside. This was partially due to the fact that the manual ground truths of the MRA segmentation primarily focus on the main arteries rather than the surface vasculatures, which is of interest in neurosurgical planning [13,14]. The misclassifications may also be caused by training the model with random patches that were mostly taken from the center of the brain. In the future, we will incorporate random patches that consider both vessel and non-vessel regions, along with an increased number of patches.

The recent work by Li et al. [15] introduced a global vascular context network (GVC-Net) with a hybrid loss to address over-segmentation issues caused by sparse labels and skull vessels. They also utilized the TubeTk dataset by training on 42 data points and testing on 10 data points, achieving a sensitivity of 61.24%, precision of 75.58%, DC of 67.66%, intersection over the union of 51.13%, and centerline Dice coefficient of 83.79%. Although their method has a 5% higher Dice coefficient, it is important to note that their reported result is based on a single fold while our method's performance was from a full 4-fold cross-validation. In a separate study, Tang et al. [3] compared their proposed RPNet approach to PANet [8] in 3D abdominal image segmentation. RPNet outperformed PANet, achieving approximately 33% higher performance. Given these promising results, it would be worthwhile to investigate RPNet's application for 3D cerebral blood vessel segmentation task.

One major benefit of few-shot learning is the capacity to allow high flexibility and adaptability for unseen classes, which can include new classification/segmentation tasks and image contrasts. Our proposed approach shows sufficient generalizability to new classes as only a few image patches in the support set allowed the segmentation of the whole brain volume. In contrast, Holroyd et al. [16] developed tUbe net, a model that achieved high performance in segmenting new blood vessels through transfer learning. However, tUbe net requires a large training dataset, which may not always be available. Our method is independent of pre-trained weights and can be potentially applicable to diverse applications.

Despite efforts to utilize limited annotated data, the current performance of VesselShot is not sufficiently accurate for clinical deployment. However, we will explore the strategies mentioned above to improve the accuracy and robustness of few-shot cerebral vascular segmentation. Despite its limitations, our proposed method represents a preliminary step in addressing limited annotated data in the challenging task of 3D cerebral blood vessel segmentation.

6   Conclusion

Our novel method utilizes few-shot learning to address the challenges of limited labeled datasets in 3D cerebral blood vessel segmentation. This approach shows promise in overcoming the bottleneck of limited manually annotated datasets and could aid in clinical tasks with further improvement in the future. While the present achievement may not yet find direct application in clinical environments, it signifies an advancement in this domain.

Acknowledgements. This study was funded by an FRQNT Team Grant (2022-PR-296459).

References

1. Goni, M.R., Ruhaiyem, N.I.R., Mustapha, M., Achuthan, A., Nassir, C.M.N.C.M.: Brain vessel segmentation using deep learning-a review. In: IEEE Access (2022)
2. Roy, A.G., Siddiqui, S., Pölsterl, S., Navab, N., Wachinger, C.: squeeze & excite'guided few-shot segmentation of volumetric images. Med. Image Anal. **59**, 101587 (2020)
3. Tang, H., Liu, X., Sun, S., Yan, X., Xie, X.: Recurrent mask refinement for few-shot medical image segmentation. In: Proceedings of the IEEE/CVF International Conference on Computer Vision, pp. 3918–3928 (2021)
4. Ouyang, C., Biffi, C., Chen, C., Kart, T., Qiu, H., Rueckert, D.: Self-supervision with superpixels: training few-shot medical image segmentation without annotation. In: Vedaldi, A., Bischof, H., Brox, T., Frahm, J.-M. (eds.) ECCV 2020. LNCS, vol. 12374, pp. 762–780. Springer, Cham (2020). https://doi.org/10.1007/978-3-030-58526-6_45
5. Mondal, A.K., Dolz, J., Desrosiers, C.: Few-shot 3d multi-modal medical image segmentation using generative adversarial learning. arXiv preprint arXiv:1810.12241 (2018)
6. Guo, S., Xu, L., Feng, C., Xiong, H., Gao, Z., Zhang, H.: Multi-level semantic adaptation for few-shot segmentation on cardiac image sequences. Med. Image Anal. **73**, 102170 (2021)
7. Xu, J., et al.: A few-shot learning-based retinal vessel segmentation method for assisting in the central serous chorioretinopathy laser surgery. Front. Med. **9**, 821565 (2022)
8. Wang, K., Liew, J.H., Zou, Y., Zhou, D., Feng, J.: PANet: few-shot image semantic segmentation with prototype alignment. In: Proceedings of the IEEE/CVF International Conference on Computer Vision, pp. 9197–9206 (2019)
9. Wang, Y., et al.: VC-Net: deep volume-composition networks for segmentation and visualization of highly sparse and noisy image data. IEEE Trans. Visual Comput. Graph. **27**(2), 1301–1311 (2020)
10. Isensee, F., Jaeger, P.F., Kohl, S.A., Petersen, J., Maier-Hein, K.H.: nnU-Net: a self-configuring method for deep learning-based biomedical image segmentation. Nat. Methods **18**(2), 203–211 (2021)
11. Su, J., et al.: DV-Net: accurate liver vessel segmentation via dense connection model with D-BCE loss function. Knowl. Based Syst. **232**, 107471 (2021)
12. Livne, M., et al.: A u-net deep learning framework for high performance vessel segmentation in patients with cerebrovascular disease. Front. Neurosci. **13**, 97 (2019)
13. Hellum, O., Mu, Y., Kersten-Oertel, M., Xiao, Y.: A novel prototype for virtual-reality-based deep brain stimulation trajectory planning using voodoo doll annotation and eye-tracking. Comput. Methods Biomech. Biomed. Eng. Imaging Visual. **10**(4), 418–424 (2022)
14. Bériault, S., et al.: Towards computer-assisted deep brain stimulation targeting with multiple active contacts. In: Ayache, N., Delingette, H., Golland, P., Mori, K. (eds.) MICCAI 2012. LNCS, vol. 7510, pp. 487–494. Springer, Heidelberg (2012). https://doi.org/10.1007/978-3-642-33415-3_60

15. Li, M., Li, S., Han, Y., Zhang, T.: GVC-Net: global vascular context network for cerebrovascular segmentation using sparse labels. IRBM **43**(6), 561–572 (2022)
16. Holroyd, N.A., Li, Z., Walsh, C., Brown, E.E., Shipley, R.J., Walker-Samuel, S.: tUbe net: a generalizable deep learning tool for 3d vessel segmentation, pp. 2023–07. bioRxiv (2023)

WaveSep: A Flexible Wavelet-Based Approach for Source Separation in Susceptibility Imaging

Zhenghan Fang[1,4], Hyeong-Geol Shin[2,3], Peter van Zijl[2,3], Xu Li[2,3], and Jeremias Sulam[1,4(✉)]

[1] Department of Biomedical Engineering, Johns Hopkins University, Baltimore, MD 21218, USA
[2] F.M. Kirby Research Center for Functional Brain Imaging, Kennedy Krieger Institute, Baltimore, MD 21205, USA
[3] Department of Radiology and Radiological Sciences, Johns Hopkins University, Baltimore, MD 21205, USA
[4] Johns Hopkins Kavli Neuroscience Discovery Institute, Baltimore, MD 21218, USA
jsulam1@jhu.edu

Abstract. The separation of signal contributions from paramagnetic and diamagnetic susceptibility sources in MRI has important implications for understanding the biological functions and health conditions of the brain. However, general and flexible deep-learning-based tools that can provide this information in humans *in vivo* are limited. For instance, the state-of-the-art deep-learning-based source separation method in quantitative susceptibility mapping (QSM) demands high-quality paramagnetic and diamagnetic maps for training and only allows phase measurement from a single head orientation as input. Furthermore, no method currently exists to separate these contributions when considering the susceptibility *anisotropy* as in the more challenging framework of susceptibility tensor imaging (STI). In this paper, we present a unified and flexible algorithm for source separation for both QSM and STI, dubbed WaveSep. Our method allows for an arbitrary number of input measurements at any head orientations for better estimation accuracy given multiple input measurements, does not require ground-truth paramagnetic and diamagnetic data for training, and is able to estimate the anisotropic second-order susceptibility tensors without requiring significant additional measurements. Our method first solves the dipole inversion problem by using state-of-the-art, off-the-shelf data-driven models based on learned proximal operators, and then separates the paramagnetic and diamagnetic sources using a Wavelet-based separation approach, without the need for retraining. Experimental results on both simulation and *in-vivo* human brain data demonstrate the superior performance of WaveSep for susceptibility source separation in QSM, and

---

Z. Fang and H.-G. Shin—Equal contribution.

Supplementary Information The online version contains supplementary material available at https://doi.org/10.1007/978-3-031-44858-4_6.

unprecedented separation results in STI. Code is available at https://github.com/ZhenghanFang/WaveSep.

Keywords: MRI · Magnetic Susceptibility · Source Separation · Inverse Problems

1 Introduction

Measuring the tissue magnetic susceptibility distribution is of great value for understanding underlying biological processes in normal organ functioning [18, 26] as well as for monitoring pathological alternation in microscopic structure [11, 15]. Susceptibility-sensitive MRI has provided a way to non-invasively image this property *in vivo* in the human brain by measuring susceptibility-induced local field variations [17,21]. In particular, quantitative susceptibility mapping (QSM) assumes tissue susceptibility is isotropic [26], and estimates a scalar susceptibility per voxel from the local field variation map. The QSM reconstruction problem, however, is ill-posed, and thus theoretically more than three phase images at different head orientations are required [5,19]. On the other hand, susceptibility tensor imaging (STI) allows to measure the *anisotropic* magnetic susceptibility in highly-organized tissue (e.g., white matter) by using a second-order tensor model at each voxel, providing useful information about tissue microstructure [17]. This makes the inversion problem even more challenging, requiring at least 6 multi-orientation phase acquisitions [14,17].

Among all the tissue sources, iron and myelin are believed to be the two major sources contributing to magnetic susceptibility contrast in the brain [24], with markedly different properties: paramagnetic and isotropic iron and diamagnetic and anisotropic myelin. Estimating the individual contributions of these two elements is of great importance to understand the function and health of the brain and study neurological disorders such as multiple sclerosis [15] and Alzheimer's Disease [25]. In addition, accurate estimation of the anisotropic susceptibility of myelin also has great value for tracking neural fiber pathways, providing an alternative to diffusion weighted imaging [14].

Unfortunately, current tools that can separate the paramagnetic and diamagnetic sources in susceptibility MRI are limited [12,23,24]. In [24], simple anatomical-based constraints are used to regularize the separation problem in QSM. Chi-sepnet[12], the state-of-the-art deep learning method for QSM source separation, uses two separate neural networks to estimate total QSM and separate the sources, respectively. Nonetheless, Chi-sepnet requires ground-truth para- and dia-magnetic maps obtained from a large number of measurements for training, and is restricted to input phase images at only one head orientation, thus being unable to benefit from more measurements at additional orientations to improve estimation accuracy. Moreover, no methods currently exist that can estimate these components in STI in humans *in vivo*.

Contributions. This paper presents a flexible and unified tool for source separation in both QSM and STI. Our method incorporates wavelet-based separation with existing deep-learning methods for susceptibility estimation in either QSM or STI, thus requiring no external data or ground-truth paramagnetic or diamagnetic maps for training. Furthermore, our method has the flexibility of using phase measurements acquired at an arbitrary number of head orientations with arbitrary angles as input, achieving more accurate source separation of QSM by leveraging more phase images. Finally, our method is the first approach for source separation in STI and achieves high-quality estimations with only a few phase acquisitions.

2 Methods

We begin by describing the problem setting, first in QSM (assuming an isotropic model), and later in STI (allowing for susceptibility anisotropy).

Source Separation in QSM. By assuming isotropic susceptibility as in QSM, the susceptibility distribution can be represented by a vector $x_{qsm} \in \mathbb{R}^n$, where n denotes the number of voxels. Letting[1] $x_{pos} \geq 0 \in \mathbb{R}^n$ and $x_{neg} \leq 0 \in \mathbb{R}^n$ be the paramagnetic (i.e., positive) and diamagnetic (negative) components of x_{qsm}, we can naturally write $x_{qsm} = x_{pos} + x_{neg}$. QSM provides an estimation of the bulk (i.e. total) susceptibility, x_{qsm}, from local field shift measurements δB calculated by the induced magnetization [5]. The relationship between δB and x_{qsm} is characterized by a dipole convolution:

$$\delta B = \mathcal{F}^{-1} D \mathcal{F} x_{qsm},$$

where $\mathcal{F}$ is the Fourier transform and $D \in \mathbb{R}^{n \times n}$ is a (diagonal) dipole kernel. The kernel is singular, however, preventing the direct inversion of this system. To circumvent this limitation, modern techniques rely on either employing multiple phase measurements at different head orientations [19], employing different regularization methods, or on restoration functions learned from data [8,13,27].

Even if x_{qsm} is estimated robustly, an additional modality is required to separate para- and dia-magnetic components. The (scaled) transverse relaxometry parameter[2], $\tilde{R}'_2 \in \mathbb{R}^n$, can be used to this end, as it provides information of the absolute value sum of positive and negative susceptibilities [24], i.e.

$$\tilde{R}'_2 = |x_{pos}| + |x_{neg}| = x_{pos} - x_{neg}.$$

Hence, by combining $\tilde{R}'_2$ with x_{qsm}, one can, in principle, solve for the components x_{pos} and x_{neg}, i.e., separating paramagnetic and diamagnetic sources from the bulk susceptibility. In practice, the ill-posedness of x_{qsm} estimation problem

---

[1] We denote entrywise inequalities by $x_{pos} \geq 0$ and $x_{neg} \leq 0$.

[2] Herein we work with the transverse relaxation scaled by the relaxometric constant between original R'_2 and susceptibility, Dr, so that $\tilde{R}'_2 = \frac{1}{Dr} R'_2$.

itself and the low signal-to-noise ratio of the transverse relaxation measurement $\tilde{R}'_2$ (relative to local field measurement δB) make accurate separation challenging, and only few attempts have been made to carry this problem. For example, chisep-MEDI [24] is a model-based QSM separation method using conventional QSM regularization. Chi-sepnet [12] proposed an end-to-end deep neural network to solve this with single-orientation data.

Source Separation in STI. Different from QSM, STI considers the anisotropy in susceptibility coming from diamagnetic myelin by modeling this property as a symmetric second-order tensor $\chi \in \mathbb{R}^{3\times3}$ at every voxel. Hence, the negative (diamagnetic) component of STI can be represented by a tensor image $\mathbf{x}_{\mathrm{neg}} \in \mathbb{R}^{6n}$ by concatenating each distinct tensor element $\chi_{11}, \chi_{12}, \chi_{13}, \chi_{22}, \chi_{23}$ and χ_{33} (note that the tensor is fully determined by 6 values, as it is symmetric). The tensor at each voxel is required to be negative semidefinite, i.e. $\chi \preceq 0$, thus diamagnetic. We write $\mathbf{x}_{\mathrm{neg}} \preceq 0$ to indicate that every tensor in each voxel of $\mathbf{x}_{\mathrm{neg}}$ is negative semidefinite. The paramagnetic component can still be represented as a scalar image x_{pos}, since it results mostly from iron content which has isotropic susceptibility, as pointed out by previous work [20, 22, 24].

In this way, we can write the bulk susceptibility tensor as $\mathbf{x}_{\mathrm{sti}} = \mathbf{I}x_{\mathrm{pos}} + \mathbf{x}_{\mathrm{neg}}$, where $\mathbf{I} \in \mathbb{R}^{6n\times n}$ maps x_{pos} to an isotropic tensor image by multiplying each voxel by the identity matrix $\mathcal{I}_3 \in \mathbb{R}^{3\times3}$, i.e.,

$$(\mathbf{I}x_{\mathrm{pos}})_v = (x_{\mathrm{pos}})_v * \mathcal{I}_3 = [(x_{\mathrm{pos}})_v, 0, 0, (x_{\mathrm{pos}})_v, 0, (x_{\mathrm{pos}})_v]^T \tag{1}$$

where $(\mathbf{I}x_{\mathrm{pos}})_v$ and $(x_{\mathrm{pos}})_v$ are the mapped tensor and original susceptibility scalar at the v-th voxel, respectively.

To estimate the bulk susceptibility tensor $\mathbf{x}_{\mathrm{sti}}$, one typically employs m measurements of the local field change at different orientations:

$$\delta B^{(i)} = \mathcal{F}^{-1}D^{(i)}\mathcal{F}\mathbf{x}_{\mathrm{sti}}, \quad i = 1, ..., m, \tag{2}$$

and the relationship between $\tilde{R}'_2$, x_{pos} and $\mathbf{x}_{\mathrm{neg}}$ in STI for each orientation i is given by $\tilde{R}'^{(i)}_2 = x_{\mathrm{pos}} - H^{(i)}\mathbf{x}_{\mathrm{neg}}$, where $H^{(i)} \in \mathbb{R}^{n\times6n}$ maps $\mathbf{x}_{\mathrm{neg}}$ to a scalar image by computing the tensor projection on the i-th orientation[3]. In this higher dimensional and more ill-posed problem, no methods currently exist to perform separation of para- and diamagnetic components of the susceptibility tensor.

2.1   WaveSep: Flexible Source Separation in Susceptibility Imaging

Our general strategy is to decouple the estimation of the susceptibility distribution (either the scalar image for QSM, or the tensor image for STI) from separation of the positive and negative components. More precisely, we will first

---

[3] I.e. $(H^{(i)}\mathbf{x}_{\mathrm{neg}})_v = h^{(i)T}(\mathbf{x}_{\mathrm{neg}})_v h^{(i)}$, where $h^{(i)} \in \mathbb{R}^3$ is a unit vector representing the direction of magnetic field in the subject frame of reference, and $(\mathbf{x}_{\mathrm{neg}})_v \in \mathbb{R}^{3\times3}$ is the tensor at the v-th voxel of $\mathbf{x}_{\mathrm{neg}}$.

estimate the bulk magnetic susceptibility, i.e., scalar x_{qsm} or tensor $\mathbf{x}_{sti}$, from the observed field shifts δB by solving the dipole inversion problem. Second, the positive and negative (semidefinite) components of these tensors will be separated given the estimated bulk susceptibility and (scaled) transverse relaxation measurements $\tilde{R}_2'$. In this way, we are able to take full advantage of modern (and well performing) methods for dipole inversion in QSM and STI based on learned proximal networks [6,13] with large flexibility, while using a separation formulation that leverages analytical (Wavelet) priors. Thus, our method is able to employ any number of head acquisitions, with arbitrary orientations, without a need for extra training of the involved models. Lastly, this approach does not require any training data, which is often time-consuming, difficult and expensive to obtain, and distribution specific. We now expand on these two steps.

A variety of machine learning-based approaches have been proposed for the estimation of bulk susceptibility in QSM [2,3,7–10,13,27]. In this paper, we adopt the learned proximal approach (LPCNN, [13]) for its flexibility to multiple head orientations and superior performance. In the more challenging case of STI, we employ the learned proximal approach from DeepSTI [6], which is the only existing method that can provide reasonable reconstruction within a clinically feasible number of measurements (e.g., less than six head orientations).

Then, given the estimated bulk susceptibility and the observed $\tilde{R}_2'$, we separate positive and negative components by solving a constrained, wavelet-based ℓ_1-regularized inverse problem. For QSM, and denoting the data-fitting terms by $f_{QSM}(x_{pos}, x_{neg}) = \frac{1}{2}\|x_{qsm} - (x_{pos} + x_{neg})\|_2^2 + \frac{1}{2}\|\tilde{R}_2' - (x_{pos} - x_{neg})\|_2^2$, we address the problem

$$\min_{x_{pos} \geq 0, x_{neg} \leq 0} f_{QSM}(x_{pos}, x_{neg}) + \rho_p\|\mathbf{W}x_{pos}\|_1 + \rho_n\|\mathbf{W}x_{neg}\|_1, \qquad (3)$$

where $\mathbf{W} \in \mathbb{R}^{n \times n}$ is an orthogonal wavelet dictionary, $\|\cdot\|_2$ and $\|\cdot\|_1$ denote the ℓ_2 and ℓ_1 norms, respectively, ρ_p and ρ_n are appropriate weights of the ℓ_1 penalties on the positive and negative parts, respectively. The term $f(x_{pos}, x_{neg})$ encourages consistency between the reconstructed images (x_{pos} and x_{neg}) and the observations x_{qsm} and $\tilde{R}_2'$. The ℓ_1 penalty terms serve as a regularization for the separation problem by encouraging sparsity under a Wavelet transform. Compared to other transformations like the Fourier transform, Wavelet transforms capture local information in both spatial and frequency domains. The Daubechies (db) wavelets [4] are known to well approximate images, preserving the edges while removing the noise. Therefore, we chose the "db4" wavelet to represent susceptibility images in this work. The constraints guarantee that x_{pos} and x_{neg} are in the positive and negative orthant, respectively.

The same general formulation applies to STI as well by extending the data-fitting terms and constraints. Given m measurements at different orientations, define $f_{STI}(x_{pos}, \mathbf{x}_{neg}) = \frac{1}{2}\|\mathbf{x}_{sti} - (\mathbf{I}x_{pos} + \mathbf{x}_{neg})\|_2^2 + \frac{1}{2m}\sum_{i=1}^{m}\|\tilde{R}_2'^{(i)} - (x_{pos} - H^{(i)}\mathbf{x}_{neg})\|_2^2$. Then, with a separable Wavelet transform, $\mathbf{W}$, for each tensor com-

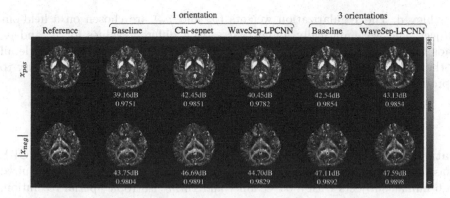

Fig. 1. QSM separation results of different methods from real *in-vivo* data. The absolute value of negative part is shown for better visualization. Reference is computed from six orientations using a COSMOS-based approach [24].

ponent of $\mathbf{x}_{\mathrm{neg}}$, we can solve

$$\min_{x_{\mathrm{pos}} \geq 0, \mathbf{x}_{\mathrm{neg}} \preceq 0} f_{\mathrm{STI}}(x_{\mathrm{pos}}, \mathbf{x}_{\mathrm{neg}}) + \rho_p \|\mathbf{W} x_{\mathrm{pos}}\|_1 + \rho_n \|\mathbf{W} \mathbf{x}_{\mathrm{neg}}\|_1. \tag{4}$$

Optimization. The problems above in Eq. (3) and Eq. (4) are convex and easily addressed by iterative proximal and projected gradient descent algorithms [1], with general form $x^{k+1} = \mathrm{Proj}_C(\mathrm{prox}_{\alpha g(\cdot)}(x^k - \alpha \nabla_x f(x^k)))$. For QSM, as an example, both para- and diamagnetic components are first updated via a gradient step of the smooth component, f_{QSM},

$$x_{pos}^{k+1} = x_{\mathrm{pos}}^k - \alpha \nabla_{x_{\mathrm{pos}}} f_{\mathrm{QSM}}(x_{\mathrm{pos}}^k, x_{\mathrm{neg}}^k) \tag{5}$$

$$x_{neg}^{k+1} = x_{\mathrm{neg}}^k - \alpha \nabla_{x_{\mathrm{neg}}} f_{\mathrm{QSM}}(x_{\mathrm{pos}}^k, x_{\mathrm{neg}}^k). \tag{6}$$

The proximal step for the ℓ_1 penalties are also simple given that we employ an orthonormal Wavelet basis, and thus this simplifies to

$$\mathrm{prox}_{\alpha \rho_p \|\mathbf{W}\cdot\|_1}(x_{\mathrm{pos}}) = \mathbf{W}^T S_{\alpha \rho_p}(\mathbf{W} x_{\mathrm{pos}}) \tag{7}$$

by means of the entry-wise soft-thresholding function $S_\beta(y) = \mathrm{sign}(y) \max(|y| - \beta, 0)$. The proximal step on x_{neg} is given similarly. Lastly, the projections onto the positive and negative orthants are trivially computed as $\mathrm{Proj}_{x \geq 0}(x_{\mathrm{pos}}) = \max(x_{\mathrm{pos}}, 0)$ and $\mathrm{Proj}_{x \leq 0}(x_{\mathrm{neg}}) = \min(x_{\mathrm{neg}}, 0)$.

The same projected proximal gradient algorithm can be employed for the STI problem as well, with analogous steps. The only difference relies on the projection step for $\mathbf{x}_{\mathrm{neg}}$ onto the negative semidefinite cone ($\mathbf{x}_{\mathrm{neg}_v} \preceq 0$). This projection can still be computed easily by means of the spectral decomposition of the (symmetric) tensor $\mathbf{x}_v = Q^T \Lambda Q$, and is given by $\mathrm{Proj}_{\mathbf{x}_v \preceq 0}(\mathbf{x}_v) = Q^T \min(\Lambda, 0) Q$.

The above iterative steps are run until the relative change of solution is lower than a predefined threshold, or a predefined maximum number of iterations is achieved. The regularization weights (ρ_p and ρ_n) are chosen on a held-out subject, and we empirically found that setting different values for ρ_p and ρ_n leads to better results in STI separation. We expand on all experimental detail in the Supplementary material, and we have released all software necessary to reproduce our experiments.

3   Experiments and Results

Data and Competing Methods. Susceptibility MRI data, including GRE phase measurements and R'_2 relaxation parameters of the brain, were acquired on 6 human subjects with a 3T scanner and $1mm^3$ isotropic spatial resolution. For each subject, measurements at six different head orientations were obtained. A COSMOS-based approach [24] is used to compute reference x_{pos} and x_{neg} maps from all six head orientations, which are considered ground truth following previous conventions [2,3,10,13]. We also generated realistic brain phantoms following steps in [6] to quantitatively evaluate the accuracy of negative semidefinite component in STI separation, where no ground truth exists for *in-vivo* data.

Table 1. Quantitative metrics of QSM and STI susceptibility source separation on real *in-vivo* data from all subjects. Note that only 1 orientation can be employed with Chi-sepnet [12].

	Number of orientations	Method	PSNR		SSIM	
			x_{pos}	x_{neg}	x_{pos}	x_{neg}
QSM	1	Baseline	39.79(0.37)	40.66(1.49)	0.975(0.003)	0.972(0.005)
		Chi-sepnet	**42.76(0.34)**	**43.31(1.45)**	**0.983(0.001)**	**0.983(0.003)**
		WaveSep-LPCNN	41.11(0.39)	41.51(1.51)	0.978(0.002)	0.975(0.004)
	2	Baseline	41.90(0.41)	42.82(1.50)	0.982(0.002)	0.980(0.004)
		WaveSep-LPCNN	**42.92(0.50)**	**43.38(1.51)**	**0.984(0.002)**	**0.982(0.003)**
	3	Baseline	43.28(0.55)	44.20(1.52)	**0.986(0.002)**	**0.984(0.003)**
		WaveSep-LPCNN	**43.87(0.56)**	**44.39(1.57)**	**0.986(0.002)**	**0.984(0.003)**
STI	3	Baseline	41.79(0.51)	39.67(1.25)	0.979(0.004)	0.963(0.005)
		WaveSep-MMSR	43.20(0.61)	43.26(1.38)	0.984(0.002)	0.981(0.003)
		WaveSep-DeepSTI	**44.48(0.57)**	**44.76(1.49)**	**0.987(0.002)**	**0.987(0.003)**
	6	Baseline	43.48(0.46)	40.45(1.25)	0.984(0.003)	0.969(0.004)
		WaveSep-MMSR	45.90(0.81)	45.58(1.18)	0.988(0.001)	0.986(0.002)
		WaveSep-DeepSTI	**47.47(0.56)**	**47.96(1.44)**	**0.992(0.001)**	**0.992(0.001)**

We compare our approach with a series of methods of increasing complexity: i) baseline approach using Wavelet-based regularization to solve dipole inversion and source separation simultaneously (similar to COSMOS-based approach but with a Wavelet-based regularization which can be applied to both QSM and STI separation), denoted as "**Baseline**", ii) a recent deep learning method for QSM separation (**Chi-sepnet** [12]), which allows only single-orientation input data, and iii) a STI separation approach using a classic STI inversion method (MMSR [16]) combined with Wavelet-based source separation (recall that no method exist for STI source separation), namely **WaveSep-MMSR**. For clarity, we denote our method by **WaveSep-LPCNN** for QSM separation and **WaveSep-DeepSTI** for STI separation.

QSM Separation Results. Figure 1 depicts visual results for QSM susceptibility source separation achieved by different methods on a representative subject using real *in-vivo* data. First, the proposed WaveSep achieves better accuracy than the baseline approach. While chi-sepnet achieves the best performance when using a single input head orientation, its network design only allows for one input measurement and thus it is incapable of benefiting from more orientations. In contrast, WaveSep, by leveraging the flexibility of incorporating more measurements, achieves better performance with increasing orientations, as expected. Table 1 summarizes the numerical results of different methods in terms of PSNR and SSIM averaged over all subjects. It can be observed that the proposed method consistently outperforms the baseline, achieves comparable results to chi-sepnet with only one additional orientation, and significantly out-

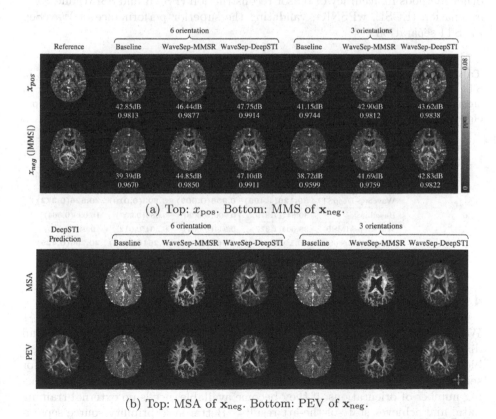

(a) Top: x_{pos}. Bottom: MMS of $\mathbf{x}_{neg}$.

(b) Top: MSA of $\mathbf{x}_{neg}$. Bottom: PEV of $\mathbf{x}_{neg}$.

Fig. 2. STI separation results from different methods on real *in-vivo* data. MMS: mean magnetic susceptibility, i.e., mean of eigenvalues of estimated tensor. MSA: magnetic susceptibility anisotropy. PEV: principal eigenvectors.

performs it with two extra orientations. Note that, unlike Chi-sepnet, WaveSep does not require any ground-truth training data for the source separation step. Therefore, WaveSep can be more readily generalized to different anatomies or cohorts where training data is lacking. We leave this interesting direction for future research.

STI Separation Results. Figure 2 shows STI separation results obtained from real *in-vivo* data, with a) x_{pos} and absolute MMS (mean magnetic susceptibility, defined as mean of eigenvalues) of the negative semidefinite tensor $\mathbf{x}_{neg}$; b) the MSA (magnetic susceptibility anisotropy, defined as $\lambda_1 - (\lambda_2 + \lambda_3)/2$ with $\lambda_1 \geq \lambda_2 \geq \lambda_3$ being the eigenvalues) and PEV (principal eigenvector) of $\mathbf{x}_{neg}$. Table 1 includes numerical metrics on real *in-vivo* data in terms of x_{pos} and MMS of $\mathbf{x}_{neg}$ (note that no ground-truth reference is available for the negative semidefinite tensor *in vivo*). To further evaluate the accuracy of $\mathbf{x}_{neg}$, we tested different methods on simulation data from brain phantoms, with numerical results outlined in Table 2. It can be observed that our method outperforms other methods in accuracy of tensor reconstruction (PSNR and SSIM) and PEV estimation (ECSE, wPSNR), validating the superior performance of WaveSep for STI separation.

Table 2. Quantitative metrics for estimation of the negative semidefinite tensor $\mathbf{x}_{neg}$ in STI separation, obtained from simulation data from realistic brain phantoms. ECSE: eigenvector cosine similarity error. wPSNR: PSNR of anisotropy-weighted principal eigenvector map.

Number of orientations	Method	PSNR	SSIM	ECSE	wPSNR
3	Baseline	30.432(1.426)	0.899(0.011)	0.408(0.006)	16.084(0.578)
	WaveSep-MMSR	36.904(1.429)	0.938(0.005)	0.362(0.008)	19.198(0.273)
	WaveSep-DeepSTI	**39.126(1.499)**	**0.956(0.003)**	**0.303(0.010)**	**20.626(0.372)**
6	Baseline	30.841(1.379)	0.900(0.008)	0.385(0.007)	16.099(0.564)
	WaveSep-MMSR	38.004(1.377)	0.944(0.004)	0.304(0.012)	19.644(0.583)
	WaveSep-DeepSTI	**40.108(1.484)**	**0.961(0.003)**	**0.261(0.013)**	**20.796(0.425)**

4   Conclusion

We presented a general and flexible framework for the separation of para- and diamagnetic components in susceptibility image of the brain, that is applicable to both QSM and STI – and in doing so, proposing the first approach for separation in STI. Our method can naturally benefit from combining an increasing number of orientations, if they become available, needs no external training data, and achieves state-of-the-art results. Better susceptibility source separation achieved by WaveSep can have critical impacts in clinical practices, enabling precise characterization of iron accumulation and demyelination in brain disorders and better diagnosis and staging of multiple sclerosis.

References

1. Beck, A.: First-Order Methods in Optimization. SIAM (2017)
2. Bollmann, S., et al.: DeepQSM - using deep learning to solve the dipole inversion for quantitative susceptibility mapping. Neuroimage **195**, 373–383 (2019)
3. Chen, Y., Jakary, A., Avadiappan, S., Hess, C.P., Lupo, J.M.: QSMGAN: improved quantitative susceptibility mapping using 3d generative adversarial networks with increased receptive field. Neuroimage **207**, 116389 (2020)
4. Daubechies, I.: Ten Lectures on Wavelets. SIAM (1992)
5. Deistung, A., Schweser, F., Reichenbach, J.R.: Overview of quantitative susceptibility mapping. NMR Biomed. **30**(4), e3569 (2017)
6. Fang, Z., Lai, K.W., van Zijl, P., Li, X., Sulam, J.: DeepSTI: towards tensor reconstruction using fewer orientations in susceptibility tensor imaging. Med. Image Anal. **87**, 102829 (2023)
7. Gao, Y., et al.: Instant tissue field and magnetic susceptibility mapping from MRI raw phase using Laplacian enhanced deep neural networks. Neuroimage **259**, 119410 (2022)
8. Jung, W., Bollmann, S., Lee, J.: Overview of quantitative susceptibility mapping using deep learning: current status, challenges and opportunities. NMR Biomed. **35**(4), e4292 (2022)
9. Jung, W., et al.: Exploring linearity of deep neural network trained QSM: QSMnet$^+$. Neuroimage **211**, 116619 (2020)
10. Kames, C., Doucette, J., Rauscher, A.: Proximal variational networks: generalizable deep networks for solving the dipole-inversion problem. In: 5th International QSM Workshop (2019)
11. Kim, H.G., et al.: Quantitative susceptibility mapping to evaluate the early stage of Alzheimer's disease. NeuroImage Clin. **16**, 429–438 (2017)
12. Kim, M., et al.: Chi-sepnet: susceptibility source separation using deep neural networks. Joint Annual Meeting ISMRM-ESMRMB & ISMRT 31st Annual Meeting, 2464 (2022)
13. Lai, K.-W., Aggarwal, M., van Zijl, P., Li, X., Sulam, J.: Learned proximal networks for quantitative susceptibility mapping. In: Martel, A.L., et al. (eds.) MICCAI 2020. LNCS, vol. 12262, pp. 125–135. Springer, Cham (2020). https://doi.org/10.1007/978-3-030-59713-9_13
14. Li, W., Liu, C., Duong, T.Q., van Zijl, P.C., Li, X.: Susceptibility tensor imaging (STI) of the brain. NMR Biomed. **30**(4), e3540 (2017)
15. Li, X., et al.: Magnetic susceptibility contrast variations in multiple sclerosis lesions. J. Magn. Reson. Imaging **43**(2), 463–473 (2016)
16. Li, X., Van Zijl, P.C.: Mean magnetic susceptibility regularized susceptibility tensor imaging (MMSR-STI) for estimating orientations of white matter fibers in human brain. Magn. Reson. Med. **72**(3), 610–619 (2014)
17. Liu, C.: Susceptibility tensor imaging. Magn. Reson. Med. Official J. Int. Soc. Magn. Reson. Med. **63**(6), 1471–1477 (2010)

18. Liu, C., Li, W., Tong, K.A., Yeom, K.W., Kuzminski, S.: Susceptibility-weighted imaging and quantitative susceptibility mapping in the brain. J. Magn. Reson. Imaging **42**(1), 23–41 (2015)

19. Liu, T., Spincemaille, P., De Rochefort, L., Kressler, B., Wang, Y.: Calculation of susceptibility through multiple orientation sampling (COSMOS): a method for conditioning the inverse problem from measured magnetic field map to susceptibility source image in MRI. Magn. Reson. Med. Official J. Int. Soc. Magn. Reson. Med. **61**(1), 196–204 (2009)

20. Oh, S.H., Kim, Y.B., Cho, Z.H., Lee, J.: Origin of b0 orientation dependent r2*(= 1/t2*) in white matter. Neuroimage **73**, 71–79 (2013)

21. Ruetten, P.P., Gillard, J.H., Graves, M.J.: Introduction to quantitative susceptibility mapping and susceptibility weighted imaging. Br. J. Radiol. **92**(1101), 20181016 (2019)

22. Schenck, J.F.: The role of magnetic susceptibility in magnetic resonance imaging: MRI magnetic compatibility of the first and second kinds. Med. Phys. **23**(6), 815–850 (1996)

23. Schweser, F., Deistung, A., Lehr, B.W., Sommer, K., Reichenbach, J.R.: Semi-twins: simultaneous extraction of myelin and iron using a t2*-weighted imaging sequence. In: Proceedings of the 19th Meeting of the International Society for Magnetic Resonance in Medicine, p. 120 (2011)

24. Shin, H.G., et al.: χ-separation: magnetic susceptibility source separation toward iron and myelin mapping in the brain. Neuroimage **240**, 118371 (2021)

25. Van Bergen, J., et al.: Colocalization of cerebral iron with amyloid beta in mild cognitive impairment. Sci. Rep. **6**(1), 1–9 (2016)

26. Wang, Y., Liu, T.: Quantitative susceptibility mapping (QSM): decoding MRI data for a tissue magnetic biomarker. Magn. Reson. Med. **73**(1), 82–101 (2015)

27. Yoon, J., et al.: Quantitative susceptibility mapping using deep neural network: QSMnet. Neuroimage **179**, 199–206 (2018)

Joint Estimation of Neural Events and Hemodynamic Response Functions from Task fMRI via Convolutional Neural Networks

Kai-Cheng Chuang[1,2(✉)] (iD), Sreekrishna Ramakrishnapillai[1,2] (iD), Krystal Kirby[3] (iD), Arend W. A. Van Gemmert[1] (iD), Lydia Bazzano[4] (iD), and Owen T. Carmichael[2] (iD)

[1] Louisiana State University, Baton Rouge, LA, USA
kchuan1@lsu.edu
[2] Pennington Biomedical Research Center, Baton Rouge, LA, USA
Owen.Carmichael@pbrc.edu
[3] Mary Bird Perkins Cancer Center, Baton Rouge, LA, USA
[4] Tulane School of Public Health and Tropical Medicine, New Orleans, LA, USA

Abstract. Joint decomposition of functional magnetic resonance imaging (fMRI) time series into time courses of neural activity events and hemodynamic response functions (HRF) can enable new insights into functional connectivity, task activation, and neurovascular coupling in health and disease. Current methods for this problem handle time series of either temporally isolated events or extended blocks of continuous events but not both; and they constrain the HRF to one specific functional form. We propose to use an autoencoder and a convolutional neural network (CNN) to overcome these challenges. The autoencoder uses convolutional neural networks to reconstruct the fMRI time series while estimating the neural event time series. The CNN estimates the HRF as the convolutional filter that, when applied to a binarized version of the neural event time series, best reconstructs the fMRI time series. When applied to synthetic data and data simulated by the STANCE fMRI simulator, the method estimates ground-truth neural events and HRFs more robustly than competing methods. When applied to real-world fMRI data, the method identifies temporally isolated, continuous, or mixed neural events that correspond to experimental conditions more closely than competing methods. The flexibility and computational power of machine learning techniques enable the accurate capture of diverse HRFs and neural event time series from fMRI data.

Keywords: HRF · Autoencoder · fMRI · CNN

Supplementary Information The online version contains supplementary material available at https://doi.org/10.1007/978-3-031-44858-4_7.

1   Introduction

Functional magnetic resonance imaging (fMRI) non-invasively records brain activity by dynamically measuring the blood oxygenation level-dependent (BOLD) signal, which reflects local changes in deoxyhemoglobin concentration in the brain and thus provides a proxy measure of neuron ensemble firing events. In the traditional BOLD signal model, the signal emerges from the convolution of neuron firing events by a hemodynamic response function (HRF), followed by the addition of noise (Fig. 1a) [1–5]. The HRF represents one aspect of neurovascular coupling, since it describes the blood oxygenation response to neural events.

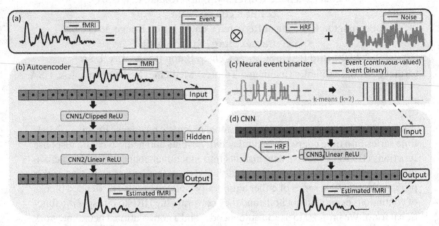

Fig. 1. **(a)** In the traditional BOLD signal model, the signal emerges from a binary neural event time series (blue) convolved by an individual- and region-specific HRF with additive noise. **(b–d)** The proposed AE-CNN architecture includes an autoencoder to estimate a continuous-valued representation of neural events **(b)**, binarization of the neural events **(c)**, and a convolutional neural network to estimate the HRF from binary neural events **(d)**. (Color figure online)

The HRF is believed to vary between individuals and between brain regions [2, 6–9], and its characteristics are believed to be valuable biomarkers of neurovascular aspects of brain health [7, 9–13]. In addition, the HRF plays a critical role in traditional task activation analyses that seek to determine which brain regions are involved in executing specific tasks [14–17]. Finally, recent work has suggested that the validity of currently popular fMRI-based functional connectivity analyses [18–22] could be improved by first deconvolving the HRF out of the fMRI signal, and assessing functional connectivity in terms of the resulting neural event time series [8, 23–26]. For all these reasons, joint estimation of the HRF and a corresponding neural event time series is of significant scientific interest.

To date, two types of solutions to this problem have been presented. The rsHRF [25] method thresholds the fMRI time series to identify neural events and then uses the neural events to estimate the parameters of an HRF that is assumed to follow a specific functional form [25]. HemoLearn and fused LASSO [11, 27, 28] estimate an HRF of a specific functional form, given a fixed neural event time series; then estimate the neural event time series, given a fixed HRF, via sparse regularization [11, 27, 28] in an alternating fashion.

This paper seeks to overcome three key limitations of these prior methods. First, prior methods excelled at estimating either *temporally isolated events* because they emphasized temporally isolated peaks in the fMRI time series (rsHRF) or *temporally continuous events* because their sparse regularizers preferred extended event periods (HemoLearn and fused LASSO). For this reason, competing methods have a difficult time handling common fMRI data sets that include both types of events [14, 15, 29, 30]. Second, two of the three prior methods were limited in their ability to handle diverse shapes of HRFs, assuming a specific parametric form to simplify estimation to a single scalar parameter [11, 27, 28]. This is an important limitation because HRF functional forms are not known a priori, vary between individuals, and vary between brain regions [31, 32]. Third, two of the three prior methods estimated continuous-valued (including positive and negative-valued) neural events [11, 27, 28], although non-binary representations of neural events are difficult to interpret in neurobiological terms [14, 15, 25, 32]. Thus, in this study, we strived to estimate binary representations, as many neuroscientists believe binary events are an intuitive and useful representation of brain functioning.

We propose to use a convolutional neural network architecture to overcome these limitations. Given the fMRI time series from a brain region of an individual, we reconstruct those time series via an autoencoder that consists of two convolutional neural network (CNN) layers and a hidden layer between them that provides a continuous-valued representation of the neural event time series (Fig. 1b). After binarizing this continuous-valued neural event time series (Fig. 1c), another CNN is trained to reconstruct the fMRI time series by convolving the binary neural event time series with a filter that the CNN estimates. This filter is the estimated HRF (Fig. 1d). The lack of mathematical constraints on the HRF allows high flexibility in HRF shape; the method is equally open to either temporally isolated or temporally continuous neural events, and the output is an intuitive binary representation of neural events. The method is fast enough that estimation of HRFs and neural events at each brain region within each scan is feasible. We show that the method identifies neural events and HRFs programmed into synthetic datasets and public-domain fMRI simulator data more accurately than competing methods. We also show that neural events detected by the method in brain regions implicated in task execution track with changes in task conditions in real-world task fMRI datasets, including datasets with temporally isolated, continuous, and mixed event designs. The HRFs estimated in the real-world data are biologically plausible.

2 Materials and Methods

2.1 Joint Estimation of Neural Events and HRF via Convolutional Neural Networks

Overview. Following the traditional BOLD signal model, we propose a convolutional neural network method that models an fMRI time series as the result of a convolution of neuron firing events with one convolution filter, the HRF, followed by the addition of noise. We use an autoencoder and a single-layer CNN (AE-CNN) to jointly model neural events and the HRF, such that the single convolutional filter of the CNN is the HRF (Fig. 1). The method consists of three steps: 1. estimation of continuous-valued neural events from input fMRI data; 2. binarization of the continuous-valued neural event representation; and 3. training a CNN whose convolutional filter represents the HRF, based on binary neural events.

Step 1: Autoencoder for Estimation of Continuous-Valued Neural Events. The autoencoder uses two layers of single-channel CNNs to reconstruct the fMRI time series in terms of a continuous-valued neural event time series that forms the hidden layer between the two CNNs. The first CNN, CNN1, generates the hidden layer via a convolutional filter and a clipped Rectified Linear Unit (clipped ReLU) activation function (threshold $= 0.3$, maximum value $= 1.5$). The second CNN, CNN2, uses another convolutional filter and a linear activation function to re-generate the input fMRI time series from the hidden layer. The CNN filters are free parameters estimated during training. CNN1 parameters are estimated with an L1 regularization penalty ($\lambda = 10^{-8}$) and no regularization penalty was applied to CNN2 parameter estimation (Fig. 1b).

Step 2: Binarization of Continuous-Valued Neural Event. Because neural events are more intuitively represented in terms of binary time series [14, 15, 25, 32], we applied k-means clustering ($k = 2$) using the squared Euclidean distance metric and the k-means $++$ algorithm [33] for cluster center initialization to all time points in the continuous-valued neural event time series and partitioned all time points into two clusters as declared binary neural events (Fig. 1c).

Step 3: CNN for Estimation of the HRF. A third single-channel CNN, CNN3, reconstructs the original fMRI time series from the binary neural event time series using a convolutional filter and a linear activation function. The convolutional filter parameters are free parameters that are estimated without a regularization penalty. The estimated convolutional filter of CNN3 comprises the HRF (Fig. 1d).

Optimization. In each CNN, the convolutional filters are the learnable parameters. The filter size of the CNNs is set depending on repetition time (TR) of the fMRI time series (28 for TR $= 1$ s, 10 for TR $= 3$ s). We use TensorFlow and Keras software packages to build our network architecture and optimize it with the Adam optimizer ($\beta 1 = 0.9$, $\beta 2 = 0.999$) with a learning rate of 0.001 to minimize a loss function of mean squared error between the reconstructed and input fMRI time series [34, 35]. Optimization takes place over a 6,000-iteration epoch in both the autoencoder and the CNN training.

Estimation of Neural Events and HRF. Each training dataset consisted of a set of fMRI time series taken from a $5 \times 5 \times 5$ cube of voxels within a specific brain region of a single fMRI scan. For each dataset, ten-fold nested cross validation (80%/20% inner split) is used to repeatedly train the neural networks, resulting in HRFs, and to quantify neural events within the validation set of the fold. The computation time for each dataset was about 20 min on an Intel Core i7 CPU. This means that the method is extensible to whole brain analysis in a tractable time frame; for example, calculating HRFs and neural events within the 120-ROI Automated Anatomical Labeling (AAL) atlas would take about 40 h. The binary neural events calculated by k-means clustering ($k = 2$) from the mean of all continuous-valued neural event time series across all ten folds of cross validation are shown in the results. The mean and standard deviation of the CNN3 filters over the ten folds are reported as the HRF. Please note that the inter-fold variability in the HRF was so small that HRF variability is not visible in Results figures.

2.2 Design of Experiments

Overview. We applied the proposed method to synthetic time series data, simulated task fMRI data from the public-domain STANCE simulator [36], and two real-world task fMRI datasets. Estimated neural event time series and HRFs were compared to those programmed into synthetic and simulated data sets. In real-world task fMRI data, neural event time series in brain regions previously implicated in task execution were compared to the time series of task conditions imposed on the participant. We generated synthetic time series with TR = 1 s, and to assess how method performance varied by repetition time, we generated simulated task fMRI time series with TR = 1 s and 3 s.

Synthetic Datasets. We generated three synthetic data sets, each representing neural events convolved by HRFs from a brain region of an individual, to test the method's ability to handle temporally isolated, temporally continuous, and temporally mixed neural events with various functional forms of the HRF. For each dataset, 125 fMRI time series were generated with the same neural event time series (Table 1). In the Temporally Isolated Dataset, a time series with 14 temporally isolated neural events was convolved with canonical HRF form 1 from the Statistical Parametric Mapping (SPM) software package [37]. In the Temporally Continuous Dataset, a time series including three blocks of temporally continuous neural events either 20 or 40 timesteps in duration was convolved with the default HRF of the HemoLearn software package [11, 28]. In the Temporally Mixed Dataset, a time series including fourteen temporally isolated neural events and two blocks of temporally continuous neural events, either 2 or 4 timesteps in duration, was convolved with canonical HRF form 2 from SPM. Each time series was 200 timesteps in duration. After convolution, uniformly distributed noise at 1% magnitude of the synthetic fMRI signal was added (i.e., a similar noise magnitude to that of real-world task fMRI data [29, 30]).

Simulated Task fMRI Datasets. For each simulated dataset, a right-handed finger related task was assumed, and 125 fMRI time series ($5 \times 5 \times 5$ voxels centered on MNI = [-38 -29 53], in the task-activated left sensorimotor cortex of an individual) were extracted from STANCE-simulated whole brain fMRI images [36]. Each time series had 200 timesteps and was convolved by a selected HRF, followed by the addition of simulated system and physiological noise at a magnitude of 1% of the simulated fMRI signal (Table 1). In the Temporally Isolated Dataset, a time series including 28 temporally isolated neural events was convolved with HRF form 1 from SPM. In the Temporally Continuous Dataset, eight blocks of temporally continuous neural events, each with 10 timesteps in duration, were embedded into the time series, which was then convolved by HRF form 3 from SPM. Temporally Mixed Dataset 1 had TR = 1s, twelve temporally isolated neural events, and two blocks of temporally continuous neural events, with each block being 3 or 7 timesteps in duration, convolved with HRF form 2 from SPM. Temporally Mixed Dataset 2 had TR = 3 s and neural events and HRF as in Temporally Mixed Dataset 1.

Table 1. Neural events, functional forms of HRF, and TR for synthetic and simulated datasets.

	Event	HRF	TR (s)
Synthetic Temporally Isolated Dataset	Temporally isolated	SPM derived form 1	1
Synthetic Temporally Continuous Dataset	Temporally continuous	HemoLearn form	1
Synthetic Temporally Mixed Dataset	Temporally mixed	SPM derived form 2	1
Simulated Temporally Isolated Dataset	Temporally isolated	SPM derived form 1	1
Simulated Temporally Continuous Dataset	Temporally continuous	SPM derived form 3	1
Simulated Temporally Mixed Dataset 1	Temporally mixed	SPM derived form 2	1
Simulated Temporally Mixed Dataset 2	Temporally mixed	SPM derived form 2	3

Real-World Task fMRI Datasets. We applied the proposed method to two in-house task fMRI datasets collected on a GE Discovery 3T scanner at Pennington Biomedical Research Center. Each task fMRI dataset had five cognitively normal older adults aged 60–85, and the fMRI time series from the task-activated brain region of each participant were trained individually. The acquisition of T1-weighted structural MPRAGE and axial 2D gradient echo EPI BOLD fMRI data (TR = 3 s) was described previously [22, 38]. Preprocessing of fMRI included slice timing correction, head motion correction, smoothing, co-registration to the T1-weighted image, and warping of T1-weighted data to Montreal Neurological Institute (MNI) coordinates. Cardiac and respiratory time series were regressed out of the data using RETROICOR [39]. In real-world dataset 1, the task consisted of five functionally discrete three-second trial types, including visual, auditory, motor, eye movement, and null trials occurring in random order [40]. Functional MRI time series ($5 \times 5 \times 5$ voxels centered on MNI = $[-24\ -2\ 50]$) were extracted from the left superior frontal gyrus region, which is expected to be activated by the motor and eye movement trials; we coded these two types of trials as binary neural events and the other trials as non-events. In real-world dataset 2, the task consisted of blocks of alternating two-finger tapping. The first block was left-handed finger tapping only; the second block required performance of the AX-continuous performance task (AX-CPT) with the right hand, and the third block included simultaneous left-handed finger tapping and right-handed AX-CPT. The details of the task design have been described previously [41]. Functional MRI time series ($5 \times 5 \times 5$ voxels centered on MNI = $[36\ -25\ 57]$) were extracted from the right sensorimotor cortex, which is expected to be activated by left-hand finger tapping. We coded both single-task and dual-task finger-tapping blocks as temporally continuous neural events and coded the rest blocks and AX-CPT blocks as non-events.

2.3 Model Evaluation

To evaluate performance, the true and false positive rates for estimated neural events were calculated as follows: *TPR = Trueestimatedevents/Totalevents* × 100%. *FPR = Falseestimatedevents/Totalnonevents* × 100%. TPR and FPR values range from 0% to 100%. A TPR of 100% corresponds to optimal model performance, while an FPR of 0% corresponds to optimal model performance. The continuous-valued neural events estimated from HemoLearn were thresholded to enable comparisons to the other methods. We selected the threshold that maximized TPR, in an attempt to present results that maximally favored HemoLearn. The R-squared between the estimated and true HRF was calculated for synthetic and simulated datasets. R-squared values range from 0 to 1, with an R-squared of one corresponding to optimal model performance.

3 Results

Synthetic Datasets. In the Temporally Isolated Dataset, the proposed method and rsHRF correctly estimated the neural event time series (TPR = 100% and FPR = 0%) and HRF (R-squared > 0.94). HemoLearn had a high true positive rate but many false positive neural events and poorly estimated the HRF (Fig. 2 and Table 2). In the Temporally Continuous Dataset, the proposed method and HemoLearn correctly estimated the neural event time series and HRF (R-squared > 0.98), but rsHRF did a poor job identifying true neural events and estimating the HRF. In the Temporally Mixed Dataset, the proposed method performed well estimating neural events (TPR = 95% and FPR = 17.22%) and the HRF (R-squared = 0.96), while rsHRF struggled in all respects, and HemoLearn had a high false neural event rate and poor HRF estimation.

Table 2. Neural event detection and HRF quantification performance for the Synthetic Datasets. The highest-performing method in each column is marked in *italics*.

	Temporally Isolated Dataset			Temporally Continuous Dataset			Temporally Mixed Dataset		
	Event		HRF	Event		HRF	Event		HRF
	TPR (%)	FPR (%)	R-squared	TPR (%)	FPR (%)	R-squared	TPR (%)	FPR (%)	R-squared
AE-CNN	*100.00*	*0.00*	*0.9997*	*100.00*	*0.00*	*0.9947*	95.00	17.22	*0.9593*
rsHRF[25]	*100.00*	*0.00*	0.9447	13.75	*0.00*	0.2857	15.00	*0.56*	0.7258
HemoLearn[11, 28]	*100.00*	85.48	0.6760	*100.00*	*0.00*	0.9849	*100.00*	78.89	0.2991

Simulated Task fMRI Datasets. In the Temporally Isolated Dataset, the proposed method correctly estimated the true neural events (TPR = 100% and FPR = 0%) and HRF (R-squared > 0.99). However, rsHRF had a low true positive rate for neural events, and HemoLearn had a high false positive rate (Table 3 and Fig. S1). In addition, both rsHRF and HemoLearn struggled to capture the HRF. In the Temporally Continuous Dataset, the proposed method and HemoLearn provided strong estimation of neural events (TPR > 97.50% and FPR < 2%) and the HRF (R-squared > 0.88). However, rsHRF had a poor neural event true positive rate and poor HRF estimation (Table 3 and

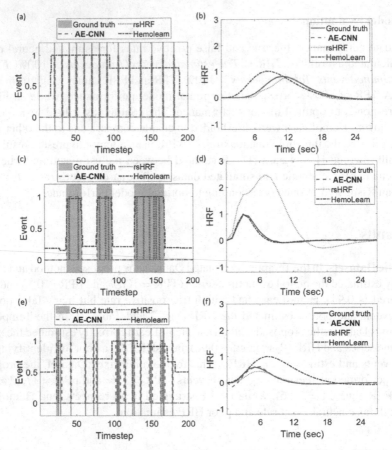

Fig. 2. Time courses of ground-truth and estimated neural events and HRF for **(a-b)** the Synthetic Temporally Isolated Dataset, **(c-d)** the Synthetic Temporally Continuous Dataset, and **(e-f)** the Synthetic Temporally Mixed Dataset.

Fig. S2). In the Temporally Mixed Dataset with TR = 1s, the proposed method estimated neural events (TPR = 100% and FPR = 3.37%) and the HRF well (R-squared = 0.92). However, rsHRF had a low true positive rate for neural events and HemoLearn had a poor false positive rate; both methods struggled to estimate the HRF (Table 3 and Fig. S3). In the Temporally Mixed Dataset with TR = 3s, the proposed method performed well in estimating neural events (TPR = 85% and FPR = 8.33%) and the HRF (R-squared = 0.86). Again, rsHRF had a low true positive rate, HemoLearn had a poor false positive rate, and both methods struggled to estimate the HRF (Fig. S4).

Real-World Task fMRI Datasets. In the Real-world Dataset 1, the proposed method estimated neural events well (TPR = 78.81 ± 6.53% and FPR = 11.70 ± 5.52% over five participants). Figure S5 shows the estimated event and HRF for a selected participant. However, rsHRF failed to identify true neural events well, and HemoLearn had a high false positive rate (Table 4). In the Real-world Dataset 2, the proposed method

Table 3. Neural event detection and HRF quantification performance for the Simulated Datasets. The highest-performing method in each column is marked in *italics*.

	Temporally Isolated Dataset			Temporally Continuous Dataset		
	Event		HRF	Event		HRF
	TPR (%)	FPR (%)	R-squared	TPR (%)	FPR (%)	R-squared
AE-CNN	*100.00*	*0.00*	*0.9997*	97.50	0.83	0.8886
rsHRF[25]	57.14	*0.00*	0.3516	10.00	*0.00*	0.5038
HemoLearn[11, 28]	96.14	73.84	0.6542	*100.00*	1.67	*0.8990*
	Temporally Mixed Dataset with TR = 1 s			Temporally Mixed Dataset with TR = 3 s		
AE-CNN	*100.00*	3.37	*0.9168*	85.00	8.33	*0.8597*
rsHRF[25]	75.00	*0.00*	0.7993	15.00	*0.56*	0.7441
HemoLearn[11, 28]	*100.00*	51.67	0.6940	75.00	20.79	0.2839

Table 4. Neural event detection performance for the two real-world data sets. The highest-performing method in each column is marked in *italics*. The mean and standard deviation over five participant data sets are reported.

	Real-world Dataset 1		Real-world Dataset 2	
	TPR (%)	FPR (%)	TPR (%)	FPR (%)
AE-CNN	78.81 ± 6.53	11.70 ± 5.52	81.83 ± 7.50	11.93 ± 1.41
rsHRF[25]	31.35 ± 6.96	*1.13 ± 0.93*	14.37 ± 2.32	*2.80 ± 0.84*
HemoLearn[11, 28]	*100.00 ± 0.00*	22.64 ± 10.72	*94.77 ± 10.19*	28.71 ± 11.73

estimated neural events well (TPR = $81.83 \pm 7.50\%$ and FPR = $11.93 \pm 1.41\%$ over five participants), and the same pattern of rsHRF and HemoLearn deficiencies were seen as in the other real data set. Figure S6 shows the estimated event and HRF for a selected participant. The median time to HRF peak across participants was 6 s, in alignment with prior studies suggesting that the typical HRF shape has a peak that occurs around three to six seconds after stimulus presentation; in addition, our HRFs show an expected post-stimulus undershoot [42, 43].

4 Conclusion

Our proposed method, AE-CNN, jointly estimated neural events and HRFs programmed into synthetic data and simulated task fMRI data. Our method outperformed the previous methods in all datasets with differing neural event structures and HRF functional forms. When applied to real-world task fMRI data, the method estimated the time course of task conditions as a proxy for neural events. Future work should extend this approach to a whole-brain approach that estimates all possible HRFs and neural events in an efficient, combined fashion rather than considering each brain region's time series independently. Extending to larger-scale real-world data and exploring the potential for HRFs to provide biomarkers for brain aging and disease are other potential avenues.

Acknowledgments. Funding for this work was provided by NIH grants R01AG041200 and R01AG062309 as well as the Pennington Biomedical Research Foundation.

References

1. Buxton, R.B.: Dynamic models of BOLD contrast. Neuroimage **62**(2), 953–961 (2012)
2. Buxton, R.B., et al.: Modeling the hemodynamic response to brain activation. Neuroimage **23**, S220–S233 (2004)
3. Buxton, R.B., Wong, E.C., Frank, L.R.: Dynamics of blood flow and oxygenation changes during brain activation: the balloon model. Magn. Reson. Med. **39**(6), 855–864 (1998)
4. Friston, K.J., Jezzard, P., Turner, R.: Analysis of functional MRI time-series. Hum. Brain Mapp. **1**(2), 153–171 (1994)
5. Friston, K.J., et al.: Nonlinear responses in fMRI: the Balloon model, Volterra kernels, and other hemodynamics. Neuroimage **12**(4), 466–477 (2000)
6. Handwerker, D.A., Ollinger, J.M., D'Esposito, M.: Variation of BOLD hemodynamic responses across subjects and brain regions and their effects on statistical analyses. Neuroimage **21**(4), 1639–1651 (2004)
7. Huettel, S.A., Singerman, J.D., McCarthy, G.: The effects of aging upon the hemodynamic response measured by functional MRI. Neuroimage **13**(1), 161–175 (2001)
8. Rangaprakash, D., et al.: Hemodynamic response function (HRF) variability confounds resting-state fMRI functional connectivity. Magn. Reson. Med. **80**(4), 1697–1713 (2018)
9. West, K.L., et al.: BOLD hemodynamic response function changes significantly with healthy aging. Neuroimage **188**, 198–207 (2019)
10. Buckner, R.L., et al.: Functional brain imaging of young, nondemented, and demented older adults. J. Cogn. Neurosci. **12**(Supplement 2), 24–34 (2000)
11. Cherkaoui, H., et al.: Multivariate semi-blind deconvolution of fMRI time series. Neuroimage **241**, 118418 (2021)
12. Rangaprakash, D., et al.: Hemodynamic variability in soldiers with trauma: implications for functional MRI connectivity studies. NeuroImage: Clin. **16**, 409–417 (2017)
13. Rangaprakash, D., et al.: FMRI hemodynamic response function (HRF) as a novel marker of brain function: applications for understanding obsessive-compulsive disorder pathology and treatment response. Brain Imaging Behav. **15**(3), 1622–1640 (2021)
14. Amaro Jr, E., Barker, G.J.: Study design in fMRI: basic principles. Brain Cogn. **60**(3), 220–232 (2006)
15. Buckner, R.L.: Event-related fMRI and the hemodynamic response. Hum. Brain Mapp. **6**(5–6), 373–377 (1998)
16. Fan, J., et al.: The activation of attentional networks. Neuroimage **26**(2), 471–479 (2005)
17. Sheu, L.K., Jennings, J.R., Gianaros, P.J.: Test–retest reliability of an fMRI paradigm for studies of cardiovascular reactivity. Psychophysiology **49**(7), 873–884 (2012)
18. Chuang, K.-C., et al.: Nonlinear conditional time-varying granger causality of task fMRI via deep stacking networks and adaptive convolutional kernels. In: Wang, L., Qi Dou, P., Fletcher, T., Speidel, S., Li, S. (eds.) Medical Image Computing and Computer Assisted Intervention – MICCAI 2022: 25th International Conference, Singapore, September 18–22, 2022, Proceedings, Part I, pp. 271–281. Springer Nature Switzerland, Cham (2022). https://doi.org/10.1007/978-3-031-16431-6_26
19. Chuang, K.-C., et al.: Deep stacking networks for conditional nonlinear granger causal modeling of fMRI data. In: Abdulkadir, A.., et al. (eds.) Machine Learning in Clinical Neuroimaging: 4th International Workshop, MLCN 2021, Held in Conjunction with MICCAI 2021, Strasbourg, France, September 27, 2021, Proceedings, pp. 113–124. Springer International Publishing, Cham (2021). https://doi.org/10.1007/978-3-030-87586-2_12

20. Friston, K., Harrison, L., Penny, W.: Dynamic causal modelling. Neuroimage **19**(4), 1273–1302 (2003)
21. Friston, K., Moran, R., Seth, A.K.: Analysing connectivity with Granger causality and dynamic causal modelling. Curr. Opin. Neurobiol. **23**(2), 172–178 (2013)
22. Chuang, K.-C., et al.: Brain effective connectivity and functional connectivity as markers of lifespan vascular exposures in middle-aged adults: the Bogalusa Heart Study. Front. Aging Neurosci. **15**, 1110434 (2023)
23. Friston, K.: Functional and effective connectivity: a review. Brain Connect. **1**(1), 13–36 (2011)
24. Wen, X., Rangarajan, G., Ding, M.: Is Granger causality a viable technique for analyzing fMRI data? PLoS ONE **8**(7), e67428 (2013)
25. Wu, G.-R., et al.: A blind deconvolution approach to recover effective connectivity brain networks from resting state fMRI data. Med. Image Anal. **17**(3), 365–374 (2013)
26. Seth, A.K., Chorley, P., Barnett, L.C.: Granger causality analysis of fMRI BOLD signals is invariant to hemodynamic convolution but not downsampling. Neuroimage **65**, 540–555 (2013)
27. Aggarwal, P., Gupta, A., Garg, A.: Joint estimation of activity signal and HRF in fMRI using fused LASSO. In: 2015 IEEE Global Conference on Signal and Information Processing (GlobalSIP). IEEE (2015)
28. Cherkaoui, H., et al.: Sparsity-based blind deconvolution of neural activation signal in fMRI. In: ICASSP 2019–2019 IEEE International Conference on Acoustics, Speech and Signal Processing (ICASSP). IEEE (2019)
29. Bühler, M., et al:, Does erotic stimulus presentation design affect brain activation patterns? Event-related vs. blocked fMRI designs. Behav. Brain Funct. **4**(1), 1–12 (2008)
30. Donaldson, D.I.: Parsing brain activity with fMRI and mixed designs: what kind of a state is neuroimaging in? Trends Neurosci. **27**(8), 442–444 (2004)
31. Asemani, D., Morsheddost, H., Shalchy, M.A.: Effects of ageing and Alzheimer disease on haemodynamic response function: a challenge for event-related fMRI. Healthcare Technol. Let. **4**(3), 109–114 (2017)
32. Glover, G.H.: Deconvolution of impulse response in event-related BOLD fMRI1. Neuroimage **9**(4), 416–429 (1999)
33. Arthur, D., Vassilvitskii, S.: K-means++ the advantages of careful seeding. In: Proceedings of the Eighteenth Annual ACM-SIAM Symposium on Discrete Algorithms (2007)
34. Abadi, M., et al.: Tensorflow: a system for large-scale machine learning. In: 12th {USENIX} Symposium on Operating Systems Design and Implementation ({OSDI} 16) (2016)
35. Chollet, F.: keras (2015)
36. Hill, J.E., et al.: A task-related and resting state realistic fMRI simulator for fMRI data validation. In: Medical Imaging 2017: Image Processing. International Society for Optics and Photonics (2017)
37. Penny, W.D., et al.: Statistical Parametric Mapping: The Analysis of Functional Brain Images. Elsevier (20110
38. Carmichael, O., et al.: High-normal adolescent fasting plasma glucose is associated with poorer midlife brain health: Bogalusa Heart Study. J. Clin. Endocrinol. Metab. **104**(10), 4492–4500 (2019)
39. Glover, G.H., Li, T.Q., Ress, D.: Image-based method for retrospective correction of physiological motion effects in fMRI: RETROICOR. Magn. Reson. Medi. Offic. J. Int. Soc. Magn. Reson. Med. **44**(1), 162–167 (2000)
40. Harvey, J.-L., et al.: A short, robust brain activation control task optimised for pharmacological fMRI studies. PeerJ **6**, e5540 (2018)
41. Kirby, K.M., et al.: Neuroimaging, behavioral, and gait correlates of fall profile in older adults. Front. Aging Neurosci. **13**, 630049 (2021)

42. Martindale, J., et al.: The hemodynamic impulse response to a single neural event. J. Cereb. Blood Flow Metab. **23**(5), 546–555 (2003)
43. Yeşilyurt, B., Uğurbil, K., Uludağ, K.: Dynamics and nonlinearities of the BOLD response at very short stimulus durations. Magn. Reson. Imaging **26**(7), 853–862 (2008)

Learning Sequential Information in Task-Based fMRI for Synthetic Data Augmentation

Jiyao Wang[1](✉), Nicha C. Dvornek[1,2], Lawrence H. Staib[1,2],
and James S. Duncan[1,2,3,4]

[1] Biomedical Engineering, Yale University, New Haven, CT 06511, USA
jiyao.wang@yale.edu
[2] Radiology and Biomedical Imaging, Yale School of Medicine, New Haven,
CT 06511, USA
[3] Electrical Engineering, Yale University, New Haven, CT 06511, USA
[4] Statistics and Data Science, Yale University, New Haven, CT 06511, USA

Abstract. Insufficiency of training data is a persistent issue in medical image analysis, especially for task-based functional magnetic resonance images (fMRI) with spatio-temporal imaging data acquired using specific cognitive tasks. In this paper, we propose an approach for generating synthetic fMRI sequences that can then be used to create augmented training datasets in downstream learning tasks. To synthesize high-resolution task-specific fMRI, we adapt the α-GAN structure, leveraging advantages of both GAN and variational autoencoder models, and propose different alternatives in aggregating temporal information. The synthetic images are evaluated from multiple perspectives including visualizations and an autism spectrum disorder (ASD) classification task. The results show that the synthetic task-based fMRI can provide effective data augmentation in learning the ASD classification task.

Keywords: Image synthesis · Data augmentation · Functional MRI · Machine learning · Medical imaging

1 Introduction

Synthetic data augmentation is a frequently used method in training machine learning models when training data is insufficient [1,5,10,14,22,24]. Although its usefulness has been demonstrated in a variety of fields related to medical imaging, most use cases are targeted towards either 2D [5,14,22] or 3D images [10] that contain only spatial information. Only a few works explore synthetic augmentation of 4D imaging data including temporal information [1,24], but fMRI is still synthesized as an individual 3D frame [24]. In this paper, we focus on augmenting the full spatio-temporal fMRI sequences from a task-based brain fMRI dataset acquired under an autism spectrum disorder (ASD) study. We show that augmenting the task-specific fMRI using an image synthesis model improves model robustness in a baseline spatio-temporal fMRI classification task. Moreover, the ability to generate synthetic fMRI data will enable fairer comparisons

A. Abdulkadir et al. (Eds.): MLCN 2023, LNCS 14312, pp. 79–88, 2023.
https://doi.org/10.1007/978-3-031-44858-4_8

of different classes of models that can be trained on the same augmented dataset, removing bias introduced by model-specific data augmentation methods.

The generative adversarial network (GAN) [6] and variational autoencoder (VAE) [9] are two popular models in image synthesis. While GANs usually suffer from disadvantages such as mode collapse and the checker-board artifact, image resolution is a challenge for VAEs. The α-GAN [10,19] architecture is a promising alternative. It modifies the GAN architecture to include auto-encoding and embedding distribution features of VAE. For our experiment, we implement an α-GAN for 4D input data to synthesize target fMRI.

In previous years, recurrent neural network structures such as long short-term memory (LSTM) [7] were frequently applied when learning sequential data. Recently, transformer structures [2,12,21], including the application of the Swin transformer in video learning [12], provide the possibility to capture long-term information in spatio-temporal data using an attention approach. Moreover, the design of the BERT [2] model highlights a potential advantage of incorporating bidirectional information in capturing sequential data. In our implementation of α-GAN, we extract spatial features from the brain using 3D convolution operations and experiment with alternatives in handling sequential spatial features including 1D convolution, LSTM, and attention.

In summary, the contributions of this work are as follows:

- We adapt the α-GAN architecture to synthesize 4D task-based fMRI data, which is to our knowledge the first to synthesize the entire spatio-temporal sequence of task-based fMRI.
- We investigate different approaches for performing temporal aggregation within the α-GAN network.
- We assess the effectiveness of fMRI image generation through quantitative analysis on brain regions related to the fMRI task, sample visualizations, and downstream use of the synthetic data in an ASD classification task.

2   Model Architecture

Following the design in Rosca et al. [19], our α-GAN model for fMRI data synthesis has four components: an encoder, a generator, a discriminator, and a code discriminator. For our application, the encoder maps a sequence of 3D volumes $X = (x_1, x_2, \ldots, x_T)$ into a compact vector embedding z. Given an embedding z and a class label L, the generator generates a 4D output X. The discriminator classifies input X between real or synthetic. The code discriminator classifies z as generated from real X or from a random standard normal distribution (Fig. 1).

Compared to a typical GAN architecture consisting of only generator and discriminator, the α-GAN model has two more components. The encoder component forms an auto-encoding structure with the generator, allowing us to utilize the reconstruction loss between the real image input to the encoder and the reconstructed output from the generator. This is especially beneficial for complex high-dimensional input data, providing the generator useful gradient information in addition to the adversarial feedback from the discriminator. It also allows us to pretrain the encoder-generator pair as an autoencoder. Meanwhile, the code discriminator component encourages the encoder-calculated embedding from real images to be similar to the embedding generated from a standard normal distribution, which is similar to the design of a VAE. Theoretically, the α-GAN model generates more stable variations in synthetic images than a typical GAN. In practice, we find that the α-GAN model also considerably improves the resolution and fineness of details in synthetic images.

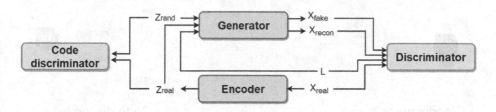

Fig. 1. α-GAN model structure

As shown in Fig. 2, we design the encoder and discriminator components in our model to process spatio-temporal information from sequential frames of fMRI. We first utilize 3D convolution to extract spatial features from each frame. Spatial features across frames are then processed by a temporal aggregation module. For the discriminator, an additional multilayer perceptron (MLP) module is included to produce the classification output. The generator component is an inverse of the encoder taking the image embedding and class label as input. Finally, the code discriminator is another MLP for classification. Here, pretrained 3D image models like ViT [3] are potential alternatives for spatial encoding. We only experiment with 3D convolution here for simplicity and focus on comparisons of temporal aggregation methods.

For the extracted temporal information, we experiment with alternatives including 1D convolution, LSTM, bidirectional LSTM, self-attention with positional encoding, and self-attention without positional encoding (Fig. 3). Theoretically, 1D convolution learns from a limited temporal kernel and shifts the same kernel along the entire sequence. It works better in capturing reoccurring local patterns. LSTM and bidirectional LSTM learn the temporal dependencies from one or both directions with a focus on remembering short-term dependencies for a long time. The attention algorithm is good at capturing long-range dependencies. When the positional encoding is removed, learning depends only on the similarity between data without considering their temporal/spatial adjacency.

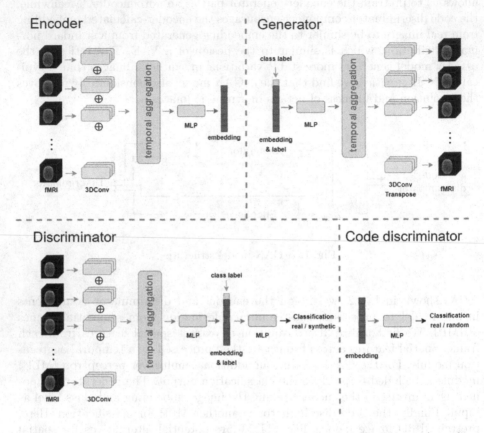

Fig. 2. α-GAN architecture diagrams for all components described above (encoder, generator, discriminator, and code discriminator)

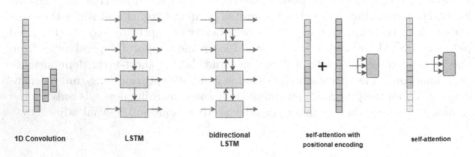

Fig. 3. Alternatives in processing temporal information applied in temporal aggregation modules of figure above

3   Data

We use a 118-subject task-based fMRI dataset acquired at the Yale Child Study Center under the "biopoint" biological motion perception task [8] designed to highlight deficits in motion perception in children with ASD. Subjects include 75 ASD children and 43 age-and-IQ-matched healthy controls. The data collection and study was approved by the Yale University Institutional Review Board (HIC #1106008625). The obtained fMRI data is preprocessed using the pipeline described in Yang et al. [23] with steps including: 1) motion correction, 2) interleaved slice timing correction, 3) BET brain extraction, 4) grand mean intensity normalization, 5) spatial smoothing, 6) high-pass temporal filtering. Each fMRI sequence contains 146 frames of $91 \times 109 \times 91$ 3D images with a frame interval of $2\,\mathrm{s}$ each. The voxel size is $3.2\,\mathrm{mm} \times 3.2\,\mathrm{mm} \times 3.2\,\mathrm{mm}$. There are 12 task stimulation videos of biological and scrambled motion, which are well aligned between subjects during the data acquisition period and given in alternating sequence. We split the dataset into 70/15/15% training/validation/test data, resulting in 72/23/23 subjects in each subset.

4   Training

During training, we apply a two-stage training scheme for the α-GAN model described above. In the first pre-training stage, the encoder and generator components are trained briefly (around 20 epochs) as an autoencoder network towards a minimum mean squared error (MSE) on 4D fMRI image reconstruction. Learned weights for both components are loaded in the second training stage to provide stable reconstruction performance at initialization. In the second training stage, training of the α-GAN model takes 3 steps including training the encoder-generator pair, discriminator, and code discriminator respectively. Let $X_{real}, X_{recon}, X_{fake}$ denote the input fMRI images, reconstructed fMRI images, and synthesized fMRI images from random embedding. z_{real} and z_{rand} denote the embedding generated from the encoder and a code sampled from the random standard normal distribution, respectively. The encoder E and generator G of our model are trained together to minimize a loss function consisting of 3 loss terms: 1) Mean absolute error (MAE) reconstruction loss between input image X_{real} and reconstructed image X_{recon}; 2) Cross entropy (CE) loss optimizing the encoder-generator pair to generate X_{recon}, X_{fake} that the discriminator classifies to be real images; 3) CE loss optimizing the encoder to generate z_{real} that the code discriminator classifies to be an image embedding generated from a random standard normal distribution. Discriminator D is trained to classify X_{real} as 1, X_{recon} and X_{fake} as 0 using CE loss. Code discriminator C is also trained using CE loss to classify z_{real} as 1, z_{rand} as 0. The losses are summarized below,

$$loss_{E,G} = \lambda ||x_{real} - x_{recon}||_1 - \log D(x_{recon}) - \log D(x_{fake}) - \log(1 - C(z_{real}))$$
$$(1)$$
$$loss_D = -\log D(x_{real}) - \log(1 - D(x_{recon})) - \log(1 - D(x_{fake})) \qquad (2)$$

$$loss_C = -\log(C(z_{real})) - \log(1 - C(z_{rand})) \tag{3}$$

where $x \in \mathbb{R}^{91 \times 109 \times 91 \times 146}$ and $z \in \mathbb{R}^{864}$. The models are implemented using PyTorch 1.10.2 [16] package and trained with the Adam optimizer under 100 epochs and a mini-batch of size 1. The learning rates for encoder-generator pair, discriminator, and code discriminator are 4, 1×10^{-6}, and 2×10^{-5} respectively. There are four consecutive 3D convolution layers for the encoder with parameters: kernel size = 16, 8, 4, 2, stride = 2, 2, 2, 1, and dimension = 4, 8, 16, 24. The generator is an inverse of the encoder using transpose convolution. The discriminator has three 3D convolution layers with parameters: kernel size = 8, 4, 4, stride = 4, 2, 1, and dimension = 4, 8, 16. For the temporal aggregation methods, 1D convolution has two layers with kernel size of 8 and stride of 4. For LSTM, we use two layers of LSTM and half the feature dimensions when changing to bidirectional. For dot-product self-attention, we use one layer of attention with raster sequence positional encoding. Training each model takes approximately 40 h on a single 40 GB Nvidia A100 GPU.

5   Evaluation and Result

First, we quantitatively analyze the similarity of fMRI signals between real data and similar-sized samples of synthetic data in three brain regions: right amygdala, fusiform gyrus, and ventromedial prefrontal cortex. These regions were identified in a previous ASD biopoint study [8] as showing salient signal changes between biological motion videos (BIO) versus scrambled motion videos (SCRAM). Ideally, the synthetic fMRI should show similar signal changes in these regions. We first use the AAL3 atlas [18] to obtain parcellations and average signals of all voxels in each region. Then, we extract fMRI sequences under SCRAM and BIO stimulation respectively and calculate the average Z-score for both sequences in Table 1. We also perform unpaired, two-tailed t-tests between signals in BIO and SCRAM frames. The p-values are listed in Table 2. The bold text in each column shows the regional pattern most similar to real fMRI. From the Z-score and t-test evaluation, the model using 1D convolution has signal most similar to real fMRI in the right amygdala and fusiform gyrus. The highest similarity in the ventromedial prefrontal cortex is achieved by the model using self-attention with positional encoding. Note that the 1D convolution model exaggerates the signal contrast between BIO and SCRAM sequences for fusiform gyrus and prefrontal cortex. Still, the 1D convolution model is the only variation that produces Z-scores with the same sign as the real fMRI for all brain regions.

To compare the distributions of real vs. synthetic fMRI sequences, we perform a tSNE [13] visualization. We generate 200 synthetic fMRI consisting of 100 synthetic ASD subjects and 100 healthy control (HC) subjects for each temporal aggregation method. Then, we apply PCA and tSNE [13] to project the 118 real and 200 synthetic fMRI onto a 3-dimensional space. See Fig. 4. All five alternatives for temporal aggregation generate synthetic data that have distribution centers similar to real fMRI in the spatio-temporal projection plots. However, considering the dispersion of data, the two plots of synthetic images generated

Table 1. Average Z-score of BIO and SCRAM Sequences

Method	Right Amygdala		Fusiform Gyrus		Prefrontal Cortex	
	SCRAM	BIO	SCRAM	BIO	SCRAM	BIO
Real fMRI	0.140	−0.144	0.165	−0.170	−0.073	0.075
1D Convolution	**0.137**	**−0.140**	**0.224**	**−0.230**	−0.219	0.225
LSTM	−0.047	0.049	0.072	−0.073	−0.061	0.063
Bidirectional LSTM	−0.089	0.091	0.029	−0.030	0.015	−0.015
Self-attention w/ PE	0.072	−0.074	−0.076	−0.078	**−0.069**	**0.071**
Self-attention w/o PE	−0.002	0.002	−0.002	0.002	−0.002	0.002

Bold text shows regional pattern most similar to real fMRI

Table 2. T-test p-value Between BIO and SCRAM Sequences

Method	Right Amygdala	Fusiform Gyrus	Prefrontal Cortex
	p-value	p-value	p-value
Real fMRI	0.089	0.043	0.374
1D Convolution	**0.095**	**0.006**	0.007
LSTM	0.565	0.385	0.858
Bidirectional LSTM	0.281	0.722	0.455
Self-attention w/ PE	0.383	0.357	**0.402**
Self-attention w/o PE	0.978	0.978	0.978

Bold text shows regional pattern most similar to real fMRI

using the attention algorithm have obviously less dispersion than the real fMRI, especially for the model trained without positional encoding. The 1D convolution result is better, while the two plots from the LSTM results have dispersion most similar to the distribution of real fMRI.

In addition to evaluations via quantitative signal analysis and tSNE embedding, we also assess the utility of the synthetic data in augmenting training data for learning an ASD versus HC classification task. The architecture of the classi-

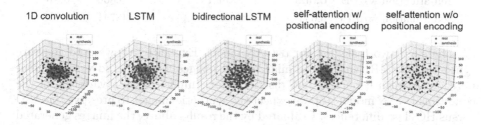

Fig. 4. Plots of tSNE projection. Each 4D fMRI is reduced to 100 dimensions by PCA and projected onto 3D by tSNE. Blue denotes real data, red denotes synthetic data. (Color figure online)

fier is shown in Fig. 5, which consists of 3D average pooling and 3D convolution operations to extract spatial features and an MLP module to calculate the classification output. The goal is to investigate the performance of synthetic fMRI for data augmentation. For classifier training, we use the 72-subject training subset of the fMRI dataset and augmented the training set to 792 samples by either adding random Gaussian noise ($\mu = 0, \sigma = 0.1$) or applying one of the five alternatives of the α-GAN model. For synthesized images using each alternative, the number of subjects in ASD and HC groups are balanced. For model selection, we save the best model evaluated by lowest validation loss during training. The resulting performances on the testing set are listed in Table 3.

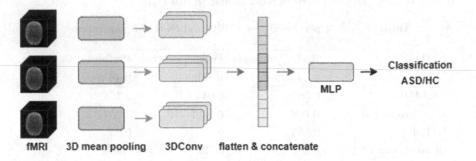

Fig. 5. Architecture of the classifier. Augmented 4D images are down-sampled spatially by mean pooling and passed to a convolutional network for an ASD classification task.

Table 3. Classifier Performances on Test Set

Method	Testing CE Loss	Balanced Acc (%)	F1 Score	AUC
w/o augmentation	0.609	68.6	0.759	0.795
Gaussian	0.731	50.0	0.686	0.697
1D Convolution	0.571	**77.7**	**0.815**	**0.833**
LSTM	0.618	73.5	0.769	0.758
Bidirectional LSTM	0.613	69.3	0.720	0.765
Self-attention w/ PE	**0.505**	**77.7**	0.800	0.814
Self-attention w/o PE	0.634	68.9	0.741	0.735

There are two tasks for our α-GAN model. Explicitly, we want to generate synthetic 4D fMRI that are similar to real images using the adversarial competition between generator and discriminator. Implicitly, as a variation of the conditional GAN model [15], we expect the synthetic images to preserve the ASD versus HC class differences. Evaluated by the results above, the images generated using 1D convolution and self-attention with positional encoding approaches have the best performance on the implicit task. Meanwhile, both approaches show noticeable improvement compared to learning from the raw dataset without augmentation.

6   Conclusion

Considering all the evaluations, 1D convolution produced the best overall performance. LSTM is usually considered a good choice for handling sequential information, but does not perform as well on our generation task. Meanwhile, the experimental results of the attention models agree with the conclusion in [20] that non-pre-trained convolutional structures are competitive and usually outperform non-pre-trained attention algorithms. Furthermore, the performance across temporal aggregation methods also enables us to make hypotheses regarding task-based fMRI data. There are two intuitive perspectives of viewing task-based fMRI, stressing either the temporal dependencies between brain states or correspondence between brain signal and task stimulation. Our results might be an indication that the task-image-correspondence plays a more important role in explaining task-based fMRI than we expected.

In recent years, various machine learning models have been applied to analyze fMRI data, including CNN, LSTM, and GNN [4,11,17]. Comparing performance within one category of models is straightforward, but comparing between categories includes bias from using different model-dependent data augmentation methods. Our method to synthesize the fMRI sequence directly removes this bias, as the same augmented dataset can be used to train all models. In the future, we intend to expand our experiments to large public datasets and apply this method as data augmentation for analysis of other fMRI data.

Acknowledgement. The data collection and study included in this paper are supported under NIH grant R01NS035193.

References

1. Abbasi-Sureshjani, S., Amirrajab, S., Lorenz, C., Weese, J., Pluim, J., Breeuwer, M.: 4D semantic cardiac magnetic resonance image synthesis on XCAT anatomical model. In: Arbel, T., Ben Ayed, I., de Bruijne, M., Descoteaux, M., Lombaert, H., Pal, C. (eds.) Proceedings of the Third Conference on Medical Imaging with Deep Learning, 06–08 Jul 2020, vol. 121, pp. 6–18. PMLR. Proceedings of Machine Learning Research (2020). http://proceedings.mlr.press/v121/abbasi-sureshjani20a.html
2. Devlin, J., Chang, M., Lee, K., Toutanova, K.: BERT: pre-training of deep bidirectional transformers for language understanding. CoRR abs/1810.04805 (2018). arXiv arxiv.org/abs/1810.04805
3. Dosovitskiy, A., et al.: An image is worth 16x16 words: transformers for image recognition at scale. CoRR abs/2010.11929 (2020). arXiv arxiv.org/abs/2010.11929
4. Dvornek, N., Ventola, P., Pelphrey, K., Duncan, J.: Identifying autism from resting-state fMRI using long short-term memory networks. In: Machine Learning in Medical Imaging, MLMI (Workshop), vol. 10541, pp. 362–370, September 2017. https://doi.org/10.1007/978-3-319-67389-9_42
5. Frid-Adar, M., Klang, E., Amitai, M., Goldberger, J., Greenspan, H.: Synthetic data augmentation using GAN for improved liver lesion classification. In: 2018 IEEE 15th International Symposium on Biomedical Imaging, ISBI 2018, pp. 289–293 (2018). https://doi.org/10.1109/ISBI.2018.8363576

6. Goodfellow, I.J., et al.: Generative adversarial networks (2014). arXiv arxiv.org/abs/1406.2661

7. Hochreiter, S., Schmidhuber, J.: Long short-term memory. Neural Comput. **9**, 1735–1780 (1997). https://doi.org/10.1162/neco.1997.9.8.1735

8. Kaiser, M.D., et al.: Neural signatures of autism. Proc. Natl. Acad. Sci. **107**(49), 21223–21228 (2010)

9. Kingma, D.P., Welling, M.: Auto-encoding variational Bayes (2013). arXiv arxiv.org/abs/1312.6114

10. Kwon, G., Han, C., Kim, D.: Generation of 3D brain MRI using auto-encoding generative adversarial networks. In: Dinggang, S., et al. (eds.) MICCAI 2019. LNCS, vol. 11766, pp. 118–126. Springer, Cham (2019). https://doi.org/10.1007/978-3-030-32248-9_14

11. Li, X., et al.: BrainGNN: interpretable brain graph neural network for fMRI analysis. Med. Image Anal. **74**, 102233 (2021)

12. Liu, Z., et al.: Video Swin transformer. arXiv arxiv.org/abs/2106.13230 (2021)

13. van der Maaten, L., Hinton, G.E.: Visualizing data using t-SNE. J. Mach. Learn. Res. **9**, 2579–2605 (2008)

14. Madan, Y., Veetil, I.K., Sowmya, V., Gopalakrishnan E.A., Soman, K.P.: Synthetic data augmentation of MRI using generative variational autoencoder for Parkinson's disease detection. In: Bhateja, V., Tang, J., Satapathy, S.C., Peer, P., Das, R. (eds.) Evolution in Computational Intelligence. Smart Innovation, Systems and Technologies, vol. 267. Springer, Singapore (2022). https://doi.org/10.1007/978-981-16-6616-2_16

15. Mirza, M., Osindero, S.: Conditional generative adversarial nets. CoRR abs/1411.1784 (2014). arXiv arxiv.org/abs/1411.1784

16. Paszke, A., et al.: Automatic differentiation in PyTorch (2017)

17. Qureshi, M.N.I., Oh, J., Lee, B.: 3D-CNN based discrimination of schizophrenia using resting-state fMRI. Artif. Intell. Med. **98**, 10–17 (2019)

18. Rolls, E.T., Huang, C.C., Lin, C.P., Feng, J., Joliot, M.: Automated anatomical labelling atlas 3. Neuroimage **206**, 116189 (2020)

19. Rosca, M., Lakshminarayanan, B., Warde-Farley, D., Mohamed, S.: Variational approaches for auto-encoding generative adversarial networks (2017). arXiv arxiv.org/abs/1706.04987

20. Tay, Y., Dehghani, M., Gupta, J.P., Bahri, D., Aribandi, V., Qin, Z., Metzler, D.: Are pre-trained convolutions better than pre-trained transformers? CoRR abs/2105.03322 (2021). arXiv arxiv.org/abs/2105.03322

21. Vaswani, A., et al.: Attention is all you need. arXiv arxiv.org/abs/1706.03762 (2017)

22. Waheed, A., Goyal, M., Gupta, D., Khanna, A., Al-Turjman, F., Pinheiro, P.R.: CovidGAN: data augmentation using auxiliary classifier GAN for improved Covid-19 detection. IEEE Access **8**, 91916–91923 (2020). https://doi.org/10.1109/ACCESS.2020.2994762

23. Yang, D., et al.: Brain responses to biological motion predict treatment outcome in young children with autism. Transl. Psychiatry **6**(11), e948 (2016). https://doi.org/10.1038/tp.2016.213

24. Zhuang, P., Schwing, A.G., Koyejo, O.: fMRI data augmentation via synthesis. In: 2019 IEEE 16th International Symposium on Biomedical Imaging, ISBI 2019, pp. 1783–1787 (2019). https://doi.org/10.1109/ISBI.2019.8759585

Clinical Applications

Causal Sensitivity Analysis for Hidden Confounding: Modeling the Sex-Specific Role of Diet on the Aging Brain

Elizabeth Haddad[1] , Myrl G. Marmarelis[2] , Talia M. Nir[1] , Aram Galstyan[2],
Greg Ver Steeg[2], and Neda Jahanshad[1,2(✉)]

[1] Imaging Genetics Center, Mark and Mary Stevens Neuroimaging and Informatics Institute,
Keck School of Medicine, University of Southern California, Los Angeles, CA, USA
njahansh@usc.edu
[2] Information Sciences Institute, University of Southern California, Los Angeles, CA, USA

Abstract. Modifiable lifestyle factors, including diet, can impact brain structure and influence dementia risk, but the extent to which diet may impact brain health for an individual is not clear. Clinical trials allow for the modification of a single variable at a time, but these may not generalize to populations due to uncaptured confounding effects. Large scale epidemiological studies can be leveraged to robustly model associations that can be specifically targeted in smaller clinical trials, while modeling confounds. Causal sensitivity analysis can be used to infer causal relationships between diet and brain structure. Here, we use a novel causal modeling approach that is robust to hidden confounding to partially identify sex-specific dose responses of diet treatment on brain structure using data from 42,032 UK Biobank participants. We find that the effects of diet on brain structure are more widespread and also robust to hidden confounds in males compared to females. Specific dietary components, including a higher consumption of whole grains, vegetables, dairy, and vegetable oils as well as a lower consumption of meat appears to be more beneficial to brain structure (e.g., greater thickness) in males. Our results shed light on sex-specific influences of hidden confounding that may be necessary to consider when tailoring effective and personalized treatment approaches to combat accelerated brain aging.

Keywords: Alzheimer's Disease · Nutrition · Lifestyle Factors

1 Introduction

In individuals at risk for Alzheimer's Disease and related diseases (ADRDs), changes in neuroimaging derived biomarkers can occur in advance of noticeable cognitive decline. Identifying risk factors associated with these changes can inform biological mechanisms, help to stratify clinical trials, and ultimately provide personalized medical advice [1]. Modifiable risk factors throughout the lifespan that can prevent or delay up to 40%

E. Haddad and M.G. Marmarelis—These authors contributed equally to this work.

of dementias worldwide have been identified and quantified in [2]. To name a few, these include less education in early life; hearing loss, brain injury and hypertension in midlife; and smoking, social isolation, and physical inactivity in late life. While diet and dietary interventions have dedicated sections in this report, quantitatively assessing their contribution to dementia risk through prodromal brain structural differences, proves challenging.

Inconsistencies in the neuroimaging correlates of nutrition may be due to heterogeneities in the way dietary habits are collected or interventions assigned (e.g., supplementation of specific vitamins/oils versus adherence to whole diets such as the Mediterranean diet), differences in study/analysis design, small samples, largely cross-sectional associations, or even due to differences in the neuroimaging measures themselves. While many studies include observed confounders as covariates (ex: age, sex, education, physical activity), it is virtually impossible to account for all factors that affect both diet treatment and brain outcomes, as complex lifestyle, environmental, and genetic interactions exist that may influence both. Also known as *hidden confounders*, these factors may be a major reason why inconsistencies across findings exist in nutritional research. Cross sectional studies generally reveal lower diet quality associated with lower brain volumes but there are also studies that fail to replicate these findings [3, 4]. Randomized control trials (RCTs) are still considered the gold standard, yet results from such trials also remain inconclusive. Some RCTs show beneficial effects of a "better quality" diet on total brain and hippocampal volumes, but not on cortical thickness, where associations have been found in some cross-sectional studies [5]. Moreover, conducting RCTs is costly, thus limiting their duration and perhaps the optimal time needed to detect noticeable changes in neuroimaging features due to dietary interventions.

Fortunately, large scale epidemiological studies that include brain imaging, like the UK Biobank [6], can now be leveraged to build robust models and make inferences that can later be specifically targeted and validated in such smaller RCTs. This can help target interventions to have more detectable effects in a shorter time frame. For example, sex is rarely investigated as a variable of interest in nutrition studies [3], even though sex and sex hormones are known to confer differential effects on glucose metabolism, insulin sensitivity, fat metabolism, adiposity, protein metabolism, and muscle protein synthesis [7]. Large-scale studies allow for models to be sex-stratified, which can inform how nutrition may differentially influence male and female brain structure, leading to more targeted treatment interventions.

While observational studies are a powerful resource to study population level effects, correlation is not sufficient to infer causality and can be subject to confounding factors [8]. Causal modeling approaches can account for confounders that are not known or modeled *a priori*. Causal inference methods for observational studies may also attempt to predict *counterfactuals*, or alternative outcomes as if they were properly randomized experiments (e.g., keeping all other variables equal, would an outcome measure change if a patient had not been a smoker). Causal sensitivity analyses attempt to answer the question of how a causal prediction might be biased if hidden confounders were affecting the observational study. We recently developed a causal sensitivity analysis method for continuous-valued exposures [9] rather than the classical binary treatment vs control setting. The use of continuous indicators reduces the degree of bias introduced by having to choose a threshold and enables the estimation of incremental effects. Identifying marginal effects of incrementally increasing exposures is perhaps more actionable than the broader proposition of completely including or eliminating an exposure. It can also be easier to identify statistically because the predictive model would make smaller extrapolations from the observed scores. Since the causal outcomes cannot be exactly identified in the presence of hidden-confounding bias, these quantities are *partially identified* using ignorance bounds that are computed using the sensitivity model.

Here, we seek to partially identify dose responses of several continuous dietary components on brain cortical thickness and subcortical volumes using a novel methodology that is robust to hidden confounding. We interrogate sex-specific models to decipher the heterogeneous dose responses of particular dietary components on regional brain structure in a large, community dwelling, aging population.

2 Methods

2.1 Partial Identification of Dose Responses

Causal inference with potential outcomes [10] is classically formulated with a dataset of triples (Y, T, X) that represent the observed outcome, assigned treatment, and covariates that include any observed confounders. The first assumption in the potential-outcomes framework is the "stable unit treatment value assumption" (SUTVA), which requires that the outcome for an individual does not depend on the treatment assignment of others. We assume that our sample of observations is independently and identically distributed, which subsumes the SUTVA. The second assumption is that of overlap, which simply means that all treatment values have a nonzero probability of occurring for all individuals. Finally, the third assumption is that of *ignorability*: $\{Y_t \perp T\}|X$. This assumption requires that there are no hidden confounders, so that the *potential* outcome Y_t is independent of the treatment assignment after conditioning on the covariates. The potential outcome is conceptually different from the observed outcome. Every individual is assumed to have a set of potential outcomes corresponding to the whole set of possible treatment assignments. The treatment variable then controls which potential outcome is actually observed: $Y = Y_T$. In this paper, we study bounded violations to the third assumption of potential outcomes by performing a causal sensitivity analysis. Causal sensitivity analysis allows one to make reasonable structural assumptions about causal effects across varying levels of hidden confounding.

Dose Responses. Treatment variables analyzed in our study were continuous diet *scores*. Formally, we study the conditional average causal derivative (CACD) defined as $\frac{\partial}{\partial t} E[Y_t|X]$ at the observed treatment, $t = T$. We present the CACD as the percent change from the observed outcome, with respect to the observed diet-score treatment variable, which takes on continuous values from 1–10.

Approach to Partial Identification. By performing a causal sensitivity analysis using a suitable *sensitivity model*, we may *partially* identify causal effects even under the influence of possible hidden confounders. Our novel marginal sensitivity model for continuous treatments (δMSM), like other sensitivity models, makes reasonable structural assumptions about the nature of possible hidden confounders on the basis of observed confounding. It differs from other sensitivity models in that it allows continuous-valued treatment variables. The δMSM exposes a single parameter, $\Gamma \geq 1$, which controls the overall amount of assumed hidden confounding. The lower and upper bounds on the estimated dose-response curves grow further apart with Γ, since it makes identification less feasible. The fundamental structural assumption for hidden confounders made by the δMSM, stated in its most compact form, is

$$\left| \frac{\partial}{\partial \tau} log \frac{p(\tau | y_t, x)}{p(\tau | x)} \right| \leq log\Gamma.$$

Here, we suppose that the logarithmic derivative of the ratio of two probability density functions is bounded. We use the notation τ to indicate the observed treatment *assignment*, which may differ from the treatment of the potential outcome that we are inquiring about. The numerator, $p(\tau | y_t, x)$, is termed the complete propensity and is unobservable. The denominator, $p(\tau | x)$, is the observable analogue termed the nominal propensity that measures association between treatment assignments and individuals as described by covariates. The complete propensity diverges from the nominal propensity only in the presence of hidden confounders.

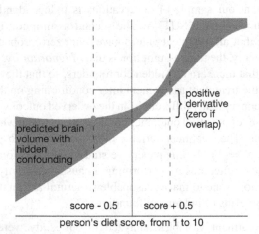

Fig. 1. Partially identified causal derivatives. For each individual, we predict outcomes after lowering a single diet score by a small amount, and also raising it by that amount. This enables approximation of the causal derivative by finite differences. If the outcomes are partially identified (where $\Gamma > 1$ so multiple outcomes and hence derivatives are admissible under the problem constraints) then we compute the absolute value of the smallest possible derivative.

It relates via Bayes' rule to a quantity that we call the counterfactual: $p(y_t|\tau, x)$. This conditional density captures the counterfactual, or potential outcome of a treatment given that the individual was *assigned* a different treatment (possibly based on confounding variables). The assignment only affects the potential outcome when the ignorability assumption is violated. In such cases, knowledge of the individual's treatment assignment $T = \tau$ is informative of one or more hidden confounders, which by definition affect the individual's intrinsic potential outcomes. What is useful for actionable causal inference is the distribution of a potential outcome conditioned just on covariates:

$$p(y_t|x) = \int p(y_t|\tau, x)p(\tau|x)d\tau,$$

which informally integrates over all the possible individuals and therefore correctly averages out the hidden confounding. Clearly, these counterfactuals cannot be inferred because the confounding influences through the treatment assignment are unobservable. In [9], we describe how to partially identify that entire integral, therefore producing bounds for admissible potential-outcome distributions and their expectations $E[Y_t|X]$. We learned two predictive models to achieve this: the observed outcome $p(y|\tau, x)$ and the nominal propensity $p(\tau|x)$. Since the diet scores are bounded within 1–10, we rescaled to the unit interval and used the Balanced Beta parametrization for the δMSM and the propensities, as described in [9]. Here, we extend our previous work to evaluate diet effects on sex-stratified models of brain structure.

Code Availability. The GitHub repository for our method may be found here: https:// github.com/marmarelis/TreatmentCurves.jl.

2.2 Analysis

Data from 42,032 UK Biobank participants (mean age ± standard deviation: 64.57 ± 7.7) with structural MRI were included. Females comprised 51.1% of this total and were on average slightly younger (64 ± 7.5) compared to males (65.2 ± 7.8).

Lifestyle, Dietary Scores, and Other Covariates. Lifestyle factors (see Table 1) documented at the time of imaging were included as covariates. Diet quality was calculated using the same coding scheme as in [11] and [12]. Briefly, an overall diet quality score was computed as the sum of individual diet components that corresponded to ideal (score of 10) and poor (score of 0) consumption of various diet components. Ideal for fruit, vegetable, whole grains, fish, dairy, and vegetable oil generally meant a higher consumption was better, whereas ideal for refined grains, processed meat, unprocessed meat, and sugary food/drink intake meant lower consumption was better. Age at scan, ApoE4 (additive coding), and Townsend deprivation index (TDI), a measure of socioeconomic status, [13] were also included as covariates.

MRI-Derived Features. Bilateral (left, right averaged) cortical thickness and subcortical volumes were derived from FreeSurfer v7.1 [14]. Cortical parcellations from the Desikan-Killiany (DK) atlas [15] were used where 34 distinct regions on each cortical hemisphere are labeled according to the gyral patterns. We also assessed 13 subcortical volumes.

Table 1. Lifestyle factors and respective UK Biobank data field IDs. American Heart Association (AHA) guidelines for weekly ideal ($\geq$150 min/week moderate or $\geq$ 75 min/week vigorous or 150 min/week mixed), intermediate (1–149 min/week moderate or 1–74 min/week vigorous or 1–149 min/week mixed), and poor (not performing any moderate or vigorous activity) physical activity were computed. Supplementation was categorized into any vitamins/minerals or fish oil intake. Salt added to food and variation in diet included the responses of "never or rarely", "sometimes", "usually", or "always/often". Coffee, tea, and water intake were integer values representing cups/glasses per day. Smoking status included never having smoked, previously smoked, and currently smokes. Alcohol frequency was categorized as infrequent (1–3 times a month, special occasions only, or never), occasional (1–2 a week or 3–4 times a week), and frequent (self-report of daily/almost daily and ICD conditions F10, G312, G621, I426, K292, K70, K860, T510). Social support/contact variables included attending any type of leisure/social group events, having family/friend visits twice a week or more, and being able to confide in someone almost daily.

Lifestyle Factor	Features (Data Field ID)
Physical Activity/ Body Composition	AHA physical activity (884, 904, 894, 914); waist to hip ratio (48,49); body mass index (BMI) (23104); body fat percentage (23099)
Sleep	sleep 7–9 h/night (1160); job involves night shift work (3426); daytime dozing/sleeping (1220)
Diet/ Supplements	diet quality scores (overall) and for the following components: fruit (1309, 1319); vegetables (1289, 1299); whole grains (1438, 1448, 1458, 1468); fish (1329, 1339); dairy (1408, 1418); vegetable oil (1428, 2654, 1438); refined grains (1438, 1448, 1458, 1468); processed meats (1349, 3680); unprocessed meats (1369, 1379, 1389, 3680); sugary foods/drinks (6144). fish oil supplementation (20084); vitamin/mineral supplementation (20084); salt added to food (1478); variation in diet (1548); water intake (1528); tea intake (1488); coffee intake (1498)
Education	college/university (6138)
Smoking	smoking status (20116)
Alcohol	alcohol intake frequency (1558/ICD)
Social Contact/Support	attending leisure/social group events (6160); frequency of friends/family visits (1031); able to confide in someone (2110)

Network Architecture. To calculate CACD, we first implemented multilayer perceptrons with single-skip connections with 40 inputs, which were the covariates and diet scores across 47 cortical and subcortical outputs. We used Swish activations in the inner layers [16] and trained with an Adam optimizer [17]. We partitioned the entire dataset into 75/25 train/test splits four times such that the test sets did not overlap and we could obtain out-of-sample predictions for all individuals. The network had the following hyperparameters: layers:3; hidden units:32; learning rate:5×10^{-3}; training epochs:10^4; batch size:$n/10$; and ensemble size:16.

Presented Outcomes. First, we present regional effect sizes for associations between 47 brain measures and diet scores derived from linear regressions, covarying for the lifestyle factors listed in Table 1. This will help serve as a comparison method given this is the most common statistical approach taken in the literature. We also present regional brain effects as causal derivatives, specifically CACDs, expressed as percentage changes from the individual's observed outcome. Therefore, the CACD can be interpreted as, "a one unit increase in diet score causes a X% increase in the outcome measure" for all measures except ventricular volumes, which represents a decrease in outcome measure. In Fig. 1, we show the CACD values closest to zero that were admissible in the set of partially identified CACDs for each individual and with each level of hidden confounding (Γ). False discovery rates [18] of all regional one-sample Student t-test p-values across males and females were represented as q. We considered a metric of robustness to hidden confounding as the largest Γ (out of a grid of tested values) for which the diminished regional CACDs survived (i.e., were nonzero) with $q < 0.05$. CACDs with effect-robustness $\Gamma \geq 1.050$ are reported as fully identified (where we assume no hidden confounding) percentage-change estimates. In the rare cases where the fully identified, $\Gamma = 1$ estimates are insignificant with $q \geq 0.05$ while also being robust to some $\Gamma > 1$, we do not present those effects.

3   Results

Model Accuracy. All model errors were ≤ 1 (z-score normalized MSE).

Cortical Thickness. The largest effect in the traditional linear model was in the transverse temporal thickness in females for average diet score, however, this region did not withstand any amount of confounding in the causal sensitivity model. In males, the superior temporal and the postcentral gyrus had relatively large effects for whole grain consumption that withstood high levels of confounding and had significant CACD. A greater extent of significant causal effects robust to hidden confounding were observed in males than females. The effects that were robust to the highest level of hidden confounding ($\Gamma = 1.10$) were observed in males in the superior temporal (CACD $= 0.042\%$; $q = 3.6 \times 10^{-80}$), postcentral (CACD $= 0.036\%$; $q = 1.1 \times 10^{-49}$), and superior frontal thickness (CACD $= 0.031\%$; $q = 2.2 \times 10^{-48}$) for whole grain intake, and the medial orbitofrontal (CACD $= 0.015\%$; $q = 8 \times 10^{-7}$) and insular thickness (CACD $= 0.009\%$; $q = 5.7 \times 10^{-5}$) for vegetable intake. Causal derivatives that withstand non-zero confounding were also observed across the brain in males for dairy, unprocessed meat (lower intake), and whole grain intake (Fig. 2).

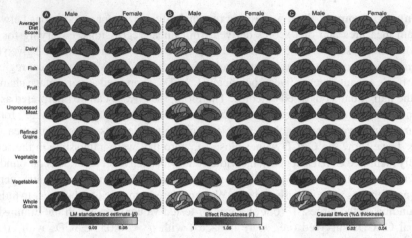

Fig. 2. A. Standardized β effects from a linear regression (LM) for thickness associations with diet scores (thresholded at uncorrected $p < 0.05$); few effects survived FDR correction including: the medial orbitofrontal for vegetables and whole grains, postcentral and superior temporal for whole grains in males, and the cuneus for fish, and transverse temporal for average diet in females. **B.** Effect robustness, or the extent of confounding that can be tolerated as causal derivatives are bounded away from zero. **C.** Causal derivatives for regions with $\Gamma \geq 1.05$ seen in B, i.e. percent difference between actual and predicted thickness per unit increase in diet score.

Subcortical Volumes. Several subcortical regions, particularly for the effects of whole grain consumption, were linearly associated with diet scores, but did not have significant causal effects in females as opposed to males. A greater extent of significant causal effects robust to hidden confounding were observed in males compared to females. The effect of the accumbens volume (CACD = 0.05%; $q = 1.4 \times 10^{-9}$) for vegetable intake was robust to the highest level of confounding ($\Gamma = 1.075$). Subcortical regions that had significant causal derivatives given a confounding level of $\Gamma = 1.05$ are highlighted in Fig. 3C.

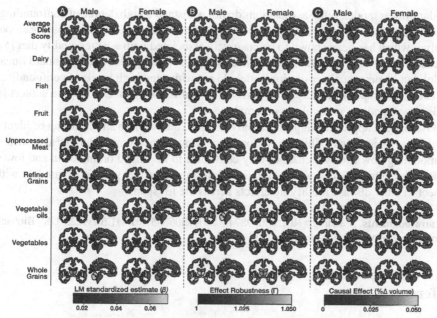

Fig. 3. A. Standardized β effects from a linear regression (LM) for subcortical volume associations with diet scores (thresholded at uncorrected $p < 0.05$); effects that survived FDR correction included: the accumbens, cerebellum cortex & white matter, and ventral DC for fruits in females; the accumbens for vegetables in males and females; the cerebellum white matter and thalamus for vegetables in males; the accumbens, amygdala, cerebellum cortex & white matter, hippocampus, thalamus, and ventral DC for whole grains in males and females; the caudate, lateral ventricle and putamen for whole grains in females; and the accumbens, amygdala, cerebellum cortex, and hippocampus in males for vegetable oil. **B.** Effect robustness, or the extent of confounding that can be tolerated as causal derivatives are bounded away from zero. **C.** Causal derivatives for regions with $\Gamma \geq 1.05$ seen in B, i.e., percent difference between actual and predicted volume per unit increase in diet score. Nucleus accumbens not pictured.

4   Discussion

Here, we use a causal sensitivity model to study the causal effects of specific diet components on brain structure from a large scale epidemiological study, as opposed to randomized control trials. Prior literature on the relationship between cortical thickness and diet show contradicting findings, where some studies fail to detect an effect while others find associations between higher thickness and adherence to a healthy diet, particularly in the entorhinal and posterior cingulate cortices [4]. A healthy diet has multiple components, and in this work, we evaluated the effect of the components of diet on male and female brain structure.

We find that the causal effects of incremental changes of a better diet appeared more robust to confounding factors in males compared to females. Our model suggests that in males, a higher intake of whole grains, vegetables, dairy, and vegetable oil, and a lower intake of unprocessed meat results in higher cortical thickness and subcortical volumes. While some of the strongest effects in our causal model were also detected in a standard linear model, other associations do not withstand a high degree of confounding, particularly in females. For example, several subcortical regions, including hippocampal volume, which has been shown to be positively associated with a better quality diet [5], are associated with a more favorable consumption of whole grains in the standard linear model, but in our casual sensitivity model, do not withstand high levels of confounding. We caution against overinterpreting cross-sectional results as these may be subject to hidden confounding, particularly in females.

Overall, our approach suggests causal effects of diet on brain structure can be identified despite some degree of hidden confounding, allowing new computational approaches for modeling how lifestyle changes may contribute to improved brain health and lower ADRD risk. Future work will continue to interrogate causal sensitivity models with respect to other disease modifying effects across the lifespan.

Acknowledgments. Funding: R01AG059874, U01AG068057, P41EB05922. UK Biobank Resource under Application Number '11559'.

References

1. Yassine, H.N., et al.: Nutrition state of science and dementia prevention: recommendations of the Nutrition for Dementia Prevention Working Group. Lancet Healthy Longev. **3**, e501–e512 (2022)
2. Livingston, G., et al.: Dementia prevention, intervention, and care: 2020 report of the Lancet Commission. Lancet **396**, 413–446 (2020)
3. Jensen, D.E.A., Leoni, V., Klein-Flügge, M.C., Ebmeier, K.P., Suri, S.: Associations of dietary markers with brain volume and connectivity: a systematic review of MRI studies. Ageing Res. Rev. **70**, 101360 (2021)
4. Drouka, A., Mamalaki, E., Karavasilis, E., Scarmeas, N., Yannakoulia, M.: Dietary and nutrient patterns and brain MRI biomarkers in dementia-free adults. Nutrients. **14** (2022). https://doi.org/10.3390/nu14112345
5. Townsend, R.F., Woodside, J.V., Prinelli, F., O'Neill, R.F., McEvoy, C.T.: Associations between dietary patterns and neuroimaging markers: a systematic review. Front. Nutr. **9**, 806006 (2022)
6. Miller, K.L., et al.: Multimodal population brain imaging in the UK Biobank prospective epidemiological study. Nat. Neurosci. **19**, 1523–1536 (2016)
7. Chen, Y., Kim, M., Paye, S., Benayoun, B.A.: Sex as a biological variable in nutrition research: from human studies to animal models. Annu. Rev. Nutr. **42**, 227–250 (2022)
8. Calude, C.S., Longo, G.: The deluge of spurious correlations in big data. Found. Sci. **22**, 595–612 (2017)
9. Marmarelis, M.G., Haddad, E., Jesson, A., Jahanshad, N., Galstyan, A., Ver Steeg, G.: Partial identification of dose responses with hidden confounders. In: The 39th Conference on Uncertainty in Artificial Intelligence (2023)
10. Rubin, D.B.: Estimating causal effects of treatments in randomized and nonrandomized studies. J. Educ. Psychol. **66**, 688–701 (1974)
11. Said, M.A., Verweij, N., van der Harst, P.: Associations of combined genetic and lifestyle risks with incident cardiovascular disease and diabetes in the UK biobank study. JAMA Cardiol. **3**, 693–702 (2018)

12. Zhuang, P., et al.: Effect of diet quality and genetic predisposition on hemoglobin A1c and Type 2 diabetes risk: gene-diet interaction analysis of 357,419 individuals. Diabetes Care **44**, 2470–2479 (2021)
13. Phillimore, P., Beattie, A., Townsend, P.: Health and deprivation. inequality and the North. Croom Helm. Health Policy, London (1988)
14. 14.Fischl, B.: FreeSurfer. Neuroimage **62**, 774–781 (2012)
15. Desikan, R.S., et al.: An automated labeling system for subdividing the human cerebral cortex on MRI scans into gyral based regions of interest. Neuroimage **31**, 968–980 (2006)
16. Ramachandran, P., Zoph, B., Le, Q.V.: Searching for Activation Functions. http://arxiv.org/abs/1710.05941 (2017)
17. Kingma, D.P., Ba, J.: Adam: a method for stochastic optimization. http://arxiv.org/abs/1412.6980 (2014)
18. Benjamini, Y., Hochberg, Y.: Controlling the false discovery rate: a practical and powerful approach to multiple testing. J. R. Stat. Soc. Series B Stat. Methodol. **57**, 289–300 (1995)

MixUp Brain-Cortical Augmentations in Self-supervised Learning

Corentin Ambroise$^{(\boxtimes)}$, Vincent Frouin, Benoit Dufumier, Edouard Duchesnay, and Antoine Grigis

Université Paris-Saclay, CEA, NeuroSpin, 91191 Gif-sur-Yvette, France
corentin.ambroise9132@gmail.com

Abstract. Learning biological markers for a specific brain pathology is often impaired by the size of the dataset. With the advent of large open datasets in the general population, new learning strategies have emerged. In particular, deep representation learning consists of training a model via pretext tasks that can be used to solve downstream clinical problems of interest. More recently, self-supervised learning provides a rich framework for learning representations by contrasting transformed samples. These methods rely on carefully designed data manipulation to create semantically similar but syntactically different samples. In parallel, domain-specific architectures such as spherical convolutional neural networks can learn from cortical brain measures in order to reveal original biomarkers. Unfortunately, only a few surface-based augmentations exist, and none of them have been applied in a self-supervised learning setting. We perform experiments on two open source datasets: Big Healthy Brain and Healthy Brain Network. We propose new augmentations for the cortical brain: baseline augmentations adapted from classical ones for training convolutional neural networks, typically on natural images, and new augmentations called MixUp. The results suggest that surface-based self-supervised learning performs comparably to supervised baselines, but generalizes better to different tasks and datasets. In addition, the learned representations are improved by the proposed MixUp augmentations. The code is available on GitHub (https://github.com/neurospin-projects/2022_cambroise_surfaugment).

Keywords: Data augmentation · Spherical convolutional neural networks · Self-supervised learning · Brain structural MRI

1 Introduction

Data-driven studies of brain pathologies are often hampered by the scarcity of available data, leading to potential failures in the discovery of statistically significant biomarkers. Key factors include recruitment of rare disease patients and

Supplementary Information The online version contains supplementary material available at https://doi.org/10.1007/978-3-031-44858-4_10.

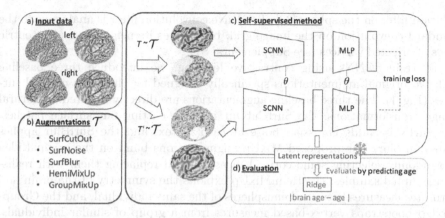

Fig. 1. Overview of the proposed evaluation framework for spherical augmentations: a) input cortical measures (here curvature) inflated to a sphere, b) a set of adapted and domain-specific augmentations T that allow the generation of augmented cortical measures, c) a self-supervised model with parameters θ consisting of a spherical Convolutional Neural Network (CNN) and a Multi-Layer Perceptron (MLP) projector, and d) evaluation of the model's frozen representations using a linear predictor (here the mean absolute deviation between linearly predicted brain age and true age).

acquisition costs. Many research efforts have attempted to address this challenge [12]. Transfer learning has become a promising solution with the advent of large cohorts such as Big Healthy Brain (BHB) [11]. Transfer learning consists of training a Neural Network (NN) with pretext tasks on a large dataset. The trained NN is then fine-tuned on a smaller, application-specific dataset. However, transfer learning for medical imaging is still in its early stages. Interestingly, there is a consensus that using a natural image dataset may not lead to the best transfer strategy [2, 21]. In recent years, several training schemes have been proposed for learning "universal" data representations [3]. The goal is to summarize as much of the semantic information as possible. Among the most promising approaches are self-supervised schemes that can provide good NN initialization for transfer learning [4, 16, 29]. Specifically, contrastive learning uses data augmentation to structure the learned latent space [5, 17, 19]. Therefore, such techniques rely heavily on the data augmentation [4, 5, 16, 28]. Currently, data augmentation for medical imaging is only available for image data defined on a regular rectangular grid.

In this work, we focus on domain-specific architectures called Spherical Convolutional Neural Networks (SCNNs). SCNNs have the potential to discover novel biomarkers from cortical brain measures. The bottleneck is the definition of convolution strategies adapted to graph or spherical topologies. Several strategies have been proposed in the literature [6, 8, 9, 18, 20, 23, 25, 32]. In neuroimaging, cortical measures are usually available on the left and right brain hemisphere surfaces of each individual. These surfaces can be inflated to icosahedral spheres [13]. Leveraging the regular and consistent geometric structure of this spherical mesh on which the cortical surface is mapped, Zhao *et al.* defined the Direct Neighbor (DiNe) convolution filter on the sphere [32]. The DiNe convolution filter is analogous to the standard convolution on the image grid, but defines its neighbors as concentric rings of oriented vertices (see Supplemental S1).

To train SCNNs using contrastive learning, we introduce three baseline and two original augmentations specifically designed for the brain cortical surfaces (Fig. 1). The three baseline augmentations are directly inspired by natural image transformations: the SurfCutOut involves cutting out surface patches, the SurfNoise adds Gaussian noise at each vertex, and the SurfBlur applies Gaussian blur. The proposed MixUp augmentations build on the original idea of randomly selecting some cortical measures and replacing them with realistic corrupted samples: the HemiMixUp exploits the symmetry of the brain and permutes measures between hemispheres of the same individual, and the Group-MixUp bootstraps vertex-based measures from a group of similar individuals. In this work, we illustrate how these augmentations can fit into the well-known SimCLR self-supervised scheme [5]. We provide a comprehensive analysis showing that self-supervised learning with SCNNs and the proposed augmentations shows similar predictive performance as supervised SCNNs for predicting age, sex and a cognitive phenotype from structural brain MRI. Furthermore, we show that the MixUp augmentations improve the learned representations and the generalization performance of the self-supervised model.

2   Methods

2.1   Baseline Brain Cortical Augmentations

All augmentations defined for natural and medical images are not directly applicable to the cortical surface. In a self-supervised scheme, an effective augmentation must reflect invariances that we want to enforce in our representation. It is not a requirement that these augmentations produce realistic samples. Their goal is to provide synthetic contrastive hard prediction tasks [5,24]. They must also be computationally efficient. For example, geometric transformations such as cropping, flipping, or jittering cannot be applied to cortical measures. We could use small rotations as proposed in [31], but such an augmentation is not computationally efficient due to multiple interpolations and less effective in nonlinear registration cases. We adapt three baseline domain-specific transformations consisting of cutting out surface patches (SurfCutOut), adding Gaussian noise at each vertex (SurfNoise), and Gaussian blurring (SurfBlur). Specifically, the SurfCutOut sets an adaptive neighborhood around a random vertex to zero. The neighborhood is defined by R concentric rings of vertices (for a definition of a ring, see the previously introduced DiNe operator). On structural MRI images, a cutout strategy has proven its efficiency in a similar contrastive learning setting [10]. Then, the SurfNoise adds a Gaussian white noise with standard deviation σ_1 (to weight the signal-to-noise ratio), and the SurfBlur smooths the data by applying a Gaussian kernel with standard deviation σ_2 (which controls the spatial extent expressed in rings).

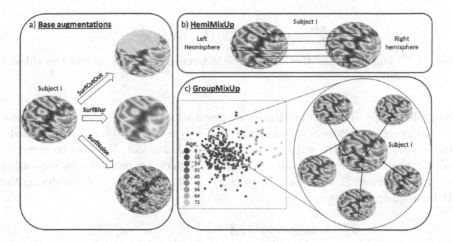

Fig. 2. Illustration of the considered cortical augmentations: a) the three baselines, b) the proposed HemiMixUp, and c) the proposed GroupMixUp brain cortical augmentations. In the GroupMixUp, groups are defined from the reduced input data using PCA embeddings colored by age. The arrows in b) and c) represent how the cortical measures are modified and are explained in more detail in Supplemental S2.

2.2 Proposed MixUp Brain Cortical Augmentations

Here, we assume that the structural measures across the cortical surface have a vertex-to-vertex correspondence for both hemispheres. We propose to randomly select vertices and their associated cortical measures and to replace them with noisy realistic ones. A similar approach has been proposed by Yoon et al. [27] for tabular data, and a comparable augmentation has been used for natural images in supervised contexts by mixing up labels [30]. All augmentations are applied on a cortical measure and hemisphere basis. A corrupted version $\tilde{\mathbf{x}}$ of a cortical measure $\mathbf{x} \in \mathbb{R}^P$, where P is the number of vertices, is generated as follows:

$$\tilde{\mathbf{x}} = \mathbf{m} \odot \bar{\mathbf{x}} + (1 - \mathbf{m}) \odot \mathbf{x} \tag{1}$$

where $\odot$ is the point-wise multiplication operator, $\bar{\mathbf{x}}$ is a noisy sample to be defined, and $\mathbf{m} \in \{0, 1\}^P$ is a binary random mask. In our case $\mathbf{x} \in \{x_1, \ldots, x_N\}$, where N is the number of subjects. The proposed HemiMixUp and GroupMixUp augmentations offer different ways to construct $\bar{\mathbf{x}}$ (Fig. 2). The mask $\mathbf{m}$ is generated by drawing binary random numbers from a Bernoulli distribution $\mathcal{B}(p)$, where p is a hyperparameter controlling the proportion of $\mathbf{x}$ to be modified. In both cases, the augmentation is done at the subject level.

HemiMixUp: This augmentation randomly permutes a subject's measurements at specific vertices across hemispheres, assuming a vertex-to-vertex correspondence between hemispheres. Considering the left hemisphere, we get:

$$\tilde{x}_i^{\text{left}} = m \odot x_i^{\text{right}} + (1 - m) \odot x_i^{\text{left}} \tag{2}$$

where x_i^{left} and x_i^{right} are the left and right hemisphere measures of the subject i, respectively.

GroupMixUp: The GroupMixUp augmentation randomly bootstraps measures at specific vertices across a group of K subjects $G_i = \{g_1, \ldots, g_K\}$ sharing similar cortical patterns with respect to the i-th subject. We aim to generate realistic noisy measures without missing hemispheric asymmetries by exploiting the group variability. We define $\mathbf{X}_i = (\mathbf{x}_{g_1}, \ldots, \mathbf{x}_{g_K})^{\text{T}} \in \mathbb{R}^{K \times P}$. Considering the left hemisphere, we get:

$$\tilde{x}_i^{\text{left}} = m \odot diag\left[MX_i^{\text{left}}\right] + (1 - m) \odot x_i^{\text{left}} \tag{3}$$

where $diag$ is the diagonal operator, and $M \in \{0, 1\}^{P \times K}$ is the random selection matrix. Each row of M selects a particular subject and is generated by drawing a random location from a uniform distribution $\mathcal{U}(1, K)$. The G_i grouping relies on a PCA trained on the residualized input data. The residualization is performed using ComBat to remove unwanted site-related noise [14]. We use K-nearest neighbors (with Euclidean distance) in the PCA space to define the group G_i. This step is performed only once before training, with little computational overhead. It is important to note that this strategy builds groups from a semantically meaningful space which maximizes the explained variance of the data. The first PCA axis is strongly related to age, as shown in Fig. 2-c. Groups are formed independently for each individual's hemisphere.

By inverting the *left* and *right* notations in Eqs. 2 and 3, the formulations hold for the right hemisphere.

3   Experiments

3.1   Data and Settings

Datasets: The T1 structural MRI data are processed with FreeSurfer, which calculates thickness, curvature, and sulcal morphology for each cortical vertex [13]. Interhemispheric registration (XHemi) is performed to obtain vertex-to-vertex mapping between hemispheres [15]. Inflated hemispheric cortical topologies are finally expressed on a regular order-5 icosahedral sphere. We use two datasets to demonstrate the proposed augmentations. First, we use the BHB, including more than 5000 individuals (age distribution 25.3 ± 15.0) coming from multiple acquisition sites [11]. We use the so-called BHB internal train and test sets. We further split the BHB internal test set into validation and test sets (hereafter referred to as internal test), preserving the population statistics (age, sex, and acquisition site). Finally, we keep unchanged the BHB external set (hereafter referred to as external test), which consists of subjects with a similar age distribution but disjoint acquisition sites. This set is used to evaluate generalization and robustness to unseen sites. Second, the Healthy Brain Network (HBN) cohort, which includes more than 1700 children (age distribution 10.95 ± 3.43) with behavioral specificities or learning problems [1]. After applying the same quality control, we keep 1407 subjects. We split the data into training (80 %) and test (20 %) sets, preserving population statistics (age, sex and acquisition site). Some subjects (1073) have the cognitive score WISC-V FSIQ available.

The Self-supervised Model: SimCLR [5] contrastive learning strategy attempts to bring two representations of the same transformed sample as close as possible in the model latent space, while repelling other representations. We implement this model following the recent literature on self-supervised learning, which consists of an encoder and a projector. For the encoder, we choose a single SCNN architecture to facilitate the comparison between the methods (see Supplemental S3). It has four convolution blocks. Each convolution block consists of a DiNe convolution layer followed by a rectified linear unit and an average pooling operator [32]. There are two branches in the first convolution block, one for each hemisphere. The resulting features are concatenated on the channel axis and piped to the network flow. For the projector, we implement the architecture recommended in [5], a Multi-Layer Perceptron (MLP). The model is trained on the BHB training set. Three augmentation combinations are considered during training: all the baseline brain cortical augmentations (Base), Base + HemiMixUp, and Base + GroupMixUp. The entire procedure is repeated three times for each combination (nine trainings in total) to obtain a standard deviation for each prediction task described below.

Model Selection and Evaluation: In self-supervised learning, the training loss, even when evaluated on a validation set, indicates convergence but does not reflect the quality of the learned representations [22,26]. As suggested in the literature to overcome this problem, we add a machine learning linear predictor on top of the encoder latent representations during training (ridge for regression and logistic for classification) (Fig. 1). We then estimate and monitor the associated prediction score from the validation set at each epoch (Mean Absolute Error (MAE) and coefficient of determination R^2 for regression and Balanced Accuracy (BAcc) for classification). This score is only used to monitor the training, leaving the simCLR training process completely unsupervised. Finally, we evaluate the trained models with the same strategy for age and sex predictions on all cohorts and for FSIQ prediction on HBN. Age, sex and FSIQ are known to be proxy measures to investigate mental health [7]. They represent features that a pre-trained self-supervised model should be able to learn and generalize to unseen data. Due to the discrepancy in age distribution and the lack of clinical variables of interest, the linear predictors are fitted to the BHB and HBN training representations. Finally, the internal and external test sets and the HBN test set are used to evaluate the prediction scores.

Table 1. Evaluation of the learned representations using a machine learning linear predictor on different BHB (a)/HBN (b) sets of data, tasks, and metrics. The proposed MixUp augmentations (HemiMixUp and GroupMixUp) are evaluated against combined baseline (Base) augmentations (SurfCutOut, SurfBlur and SurfNoise) using an unsupervised SimCLR-SCNN framework. The results are compared to supervised SCNNs trained to predict either age or sex (see section **Model Comparison** for details).

a)

SimCLR-SCNN	BHB internal test			BHB external test		
Augmentations	Age		Sex	Age		Sex
	MAE ($\downarrow$)	R^2 ($\uparrow$)	BAcc ($\uparrow$)	MAE($\downarrow$)	R^2 ($\uparrow$)	BAcc ($\uparrow$)
Base	$4.87_{\pm0.14}$	$0.81_{\pm0.01}$	$0.81_{\pm0.01}$	$5.89_{\pm0.17}$	$0.50_{\pm0.03}$	$0.71_{\pm0.01}$
Base + HemiMixUp	$4.72_{\pm0.16}$	$0.83_{\pm0.01}$	$0.81_{\pm0.01}$	$5.62_{\pm0.13}$	$0.55_{\pm0.03}$	$0.71_{\pm0.02}$
Base + GroupMixUp	$\mathbf{4.55}_{\pm0.07}$	$\mathbf{0.84}_{\pm0.01}$	$\mathbf{0.82}_{\pm0.01}$	$\mathbf{5.47}_{\pm0.15}$	$\mathbf{0.58}_{\pm0.02}$	$\mathbf{0.74}_{\pm0.01}$
Supervised-SCNN						
Age-supervised	$4.00_{\pm0.12}$	$0.84_{\pm0.01}$	$0.67_{\pm0.01}$	$5.06_{\pm0.19}$	$0.61_{\pm0.02}$	$0.55_{\pm0.01}$
Sex-supervised	$6.20_{\pm0.20}$	$0.69_{\pm0.02}$	$0.86_{\pm0.01}$	$6.40_{\pm0.32}$	$0.42_{\pm0.06}$	$0.68_{\pm0.01}$

b)

SimCLR-SCNN	HBN test				
Augmentations	Age		Sex	FSIQ	
	MAE ($\downarrow$)	R^2 ($\uparrow$)	BAcc ($\uparrow$)	MAE($\downarrow$)	R^2 ($\uparrow$)
Base	$1.69_{\pm0.07}$	$0.65_{\pm0.02}$	$0.80_{\pm0.01}$	$12.97_{\pm0.25}$	$0.10_{\pm0.02}$
Base + HemiMixUp	$1.68_{\pm0.02}$	$\mathbf{0.66}_{\pm0.01}$	$0.80_{\pm0.01}$	$\mathbf{12.71}_{\pm0.16}$	$0.13_{\pm0.02}$
Base + GroupMixUp	$\mathbf{1.66}_{\pm0.02}$	$\mathbf{0.66}_{\pm0.01}$	$\mathbf{0.81}_{\pm0.01}$	$12.60_{\pm0.29}$	$0.14_{\pm0.03}$
Supervised-SCNN					
Age-supervised	$1.74_{\pm0.05}$	$0.63_{\pm0.02}$	$0.66_{\pm0.01}$	$13.05_{\pm0.06}$	$0.09_{\pm0.002}$
Sex-supervised	$1.72_{\pm0.01}$	$0.63_{\pm0.01}$	$0.82_{\pm0.0048}$	$12.79_{\pm0.01}$	$0.09_{\pm0.01}$

Model Comparison: We compare the proposed SimCLR-SCNN model with supervised SCNNs. The supervised models consist of the same encoder followed by a linear predictor. The training loss for these supervised models depends on the task at hand (L1 for regression and cross-entropy for classification). These supervised models will be referred to as age-supervised if they were trained to predict age, and sex-supervised if they were trained to predict sex. Each supervised model is trained 3 times as well, to derive standard deviations, on the same train set as self-supervised models. They are evaluated the same way as self-supervised models: task-dependent machine learning linear predictors (ridge for regression and logistic for classification) are fitted to the learned representations from the SCNN encoder and evaluated on the test representations.

3.2   Results

Self-supervised SCNNs Generalize Better Than Supervised SCNNs: On BHB, compared to a much more specialized supervised SCNN setup, a

SCNN trained with the SimCLR self-supervised learning framework and the proposed augmentations shows a rather comparable performance for each of the investigated tasks (Table 1-a). For example, on the internal test, the SimCLR-SCNN age-MAE scores are 4.87, 4.72, and 4.55 for the Base, Base + HemiMixUp and Base + GroupMixUp augmentations, respectively. These scores are slighlty worse than the predictions of the age-supervised SCNN (4.0 age-MAE), which is expected since the latter was trained in a supervised manner to learn good representations for predicting age. However they remain comparable and largely outperform the sex-supervised SCNN (6.2 age-MAE). The same trend can be observed for the R^2 for the age prediction task and the BAcc for the sex prediction task for all test sets. Remarkably, for some tasks, the SimCLR-SCNN with the proposed augmentations even outperforms supervised SCNN models in terms of generalization performance. This can be seen by comparing the results on BHB internal and external test sets. For example SimCLR-SCNN loses 10% (0.81 → 0.71), 10% (0.81 → 0.71), and 8% (0.82 → 0.74) BAcc for the Base, Base + HemiMixUp and Base + GroupMixUp augmentations, between internal and external test sets, while age- and sex-supervised SCNNs lose 12% (0.67 → 0.55) and 18% (0.86 → 0.68) BAcc respectively. SimCLR SCNNs even outperform sex-supervised SCNNs for the sex prediction task on the external test set. As expected, the prediction on the external test decreases for both strategies. When the learned knowledge is transferred to HBN, we show better (or at least equivalent for sex) prediction performance for the SimCLR-SCNN with the proposed augmentations compared to the supervised SCNNs. Note that the age distribution in HBN is much narrower with a younger population than in BHB. Therefore, the MAE between Tables 1-a and 1-b cannot be directly compared. Interestingly, the R^2 values are still comparable and show a stable goodness of fit for the SimCLR-SCNN model. Note that using a supervised MLP with more than 120M parameters to predict age from the same input data gives only slightly better results than using the SimCLR-SCNN (∼2M parameters) trained with the Base + GroupMixUp augmentations (4.85 *vs* 5.47 age-MAE on the BHB external test) [11]. This suggests that the SimCLR-SCNN model with the proposed augmentations is able to learn good representations without supervision and without being too much biased by the acquisition site.

The MixUp Augmentations Improve Performance: It is clear that the MixUp augmentations improve the learned representations for each prediction task of the BHB internal and external tests (Table 1-a). In practice, we found that the GroupMixUp works better than the HemiMixUp augmentation strategy. This can be explained by the attenuation of some properties of the inter-hemispheric asymmetry forced by the HemiMixUp augmentation. Although the improvement in predicting age and sex on HBN test set is inconclusive, it is clear that HemiMixUp and GroupMixUp help in predicting FISQ, especially when looking at the R^2 metric (Table 1-b).

4 Conclusion

This study introduces cortical surface augmentations designed for training a SCNN in a self-supervised learning setup. The investigated SimCLR-SCNN shows the ability to generate representations with strong generalization properties. In fact, the learned representations from data collected from multiple sites offer promising performance, sometimes even outperforming supervised approaches. In particular, the GroupMixUp augmentation shows potential for learning stable representations across different cohorts. An ablation study could be used to further investigate the characteristics required for cortical-based data augmentation. Future work aims to incorporate prior information, such as clinical scores, into the GroupMixUp augmentation when computing the groups G_i. A similar strategy is proposed for structuring the learned representations by adding a regularization term in the training loss [10].

Data Use Declaration and Acknowledgment. The datasets analyzed during the current study are available online: OpenBHB in IEEEDataPort (doi 10.21227/7jsg-jx57), and HBN in NITRC (Release 10).

References

1. Alexander, L.M., et al.: An open resource for transdiagnostic research in pediatric mental health and learning disorders. Sci. Data **4**, 170181 (2017)
2. Alzubaidi, L., et al.: Towards a better understanding of transfer learning for medical imaging: a case study. Appl. Sci. **10**(13), 4523 (2020)
3. Bommasani, R., et al.: On the opportunities and risks of foundation models (2021)
4. Caron, M., et al.: Emerging properties in self-supervised vision transformers. In: ICCV (2021)
5. Chen, T., Kornblith, S., Norouzi, M., Hinton, G.: A simple framework for contrastive learning of visual representations. In: ICML, pp. 1597–1607 (2020)
6. Cohen, T.S., Geiger, M., Köhler, J., Welling, M.: Spherical CNNs. In: ICLR (2018)
7. Dadi, K., Varoquaux, G., Houenou, J., Bzdok, D., Thirion, B., Engemann, D.: Population modeling with machine learning can enhance measures of mental health. GigaScience **10**(10), giab071 (2021)
8. Defferrard, M., Bresson, X., Vandergheynst, P.: Convolutional neural networks on graphs with fast localized spectral filtering. In: NeurIPS, vol. 29 (2016)
9. Defferrard, M., Milani, M., Gusset, F., Perraudin, N.: DeepSphere: a graph-based spherical CNN. In: ICLR (2020)
10. Dufumier, B., et al.: Contrastive learning with continuous proxy meta-data for 3D MRI classification. In: de Bruijne, M., et al. (eds.) MICCAI 2021. LNCS, vol. 12902, pp. 58–68. Springer, Cham (2021). https://doi.org/10.1007/978-3-030-87196-3_6
11. Dufumier, B., Grigis, A., Victor, J., Ambroise, C., Frouin, V., Duchesnay, E.: OpenBHB: a large-scale multi-site brain MRI data-set for age prediction and debiasing. Neuroimage **263**, 119637 (2022)
12. Eitel, F., Schulz, M.A., Seiler, M., Walter, H., Ritter, K.: Promises and pitfalls of deep neural networks in neuroimaging-based psychiatric research. Exp. Neurol. **339**, 113608 (2021)

13. Fischl, B., Sereno, M.I., Dale, A.M.: Cortical surface-based analysis: II: inflation, flattening, and a surface-based coordinate system. Neuroimage 9(2), 195–207 (1999)
14. Fortin, J.P., et al.: Harmonization of multi-site diffusion tensor imaging data. Neuroimage **161**, 149–170 (2017)
15. Greve, D.N., et al.: A surface-based analysis of language lateralization and cortical asymmetry. J. Cogn. Neurosci. **25**(9), 1477–1492 (2013)
16. Grill, J.B., et al.: Bootstrap your own latent - a new approach to self-supervised learning. In: NeurIPS, vol. 33 (2020)
17. He, K., Fan, H., Wu, Y., Xie, S., Girshick, R.: Momentum contrast for unsupervised visual representation learning. In: CVPR (2020)
18. Jiang, C.M., Huang, J., Kashinath, K., Prabhat, Marcus, P., Niessner, M.: Spherical CNNs on unstructured grids. In: ICLR (2019)
19. Khosla, P., et al.: Supervised contrastive learning. In: NeurIPS (2020)
20. Kipf, T.N., Welling, M.: Semi-supervised classification with graph convolutional networks. In: ICLR (2017)
21. Raghu, M., Zhang, C., Kleinberg, J., Bengio, S.: Transfusion: understanding transfer learning for medical imaging. In: NeurIPS, vol. 32 (2019)
22. Saunshi, N., et al.: Understanding contrastive learning requires incorporating inductive biases. In: ICML, vol. 162, pp. 19250–19286 (2022)
23. Seong, S.B., Pae, C., Park, H.J.: Geometric convolutional neural network for analyzing surface-based neuroimaging data. Front. Neuroinform. **12**, 42 (2018)
24. Tian, Y., Sun, C., Poole, B., Krishnan, D., Schmid, C., Isola, P.: What makes for good views for contrastive learning? In: Advances in Neural Information Processing Systems, vol. 33, pp. 6827–6839. Curran Associates, Inc. (2020)
25. Veličković, P., Cucurull, G., Casanova, A., Romero, A., Lió, P., Bengio, Y.: Graph attention networks. In: ICLR (2018)
26. Wang, F., Liu, H.: Understanding the behaviour of contrastive loss. In: IEEE Conference on Computer Vision and Pattern Recognition, CVPR 2021, virtual, 19–25 June 2021, pp. 2495–2504 (2021)
27. Yoon, J., Zhang, Y., Jordon, J., van der Schaar, M.: VIME: extending the success of self- and semi-supervised learning to tabular domain. In: NeurIPS, vol. 33 (2020)
28. Zbontar, J., Jing, L., Misra, I., LeCun, Y., Deny, S.: Barlow Twins: self-supervised learning via redundancy reduction. In: ICML, vol. 139 (2021)
29. Zhai, X., Oliver, A., Kolesnikov, A., Beyer, L.: S4L: self-supervised semi-supervised learning. In: ICCV (2019)
30. Zhang, H., Cissé, M., Dauphin, Y.N., Lopez-Paz, D.: mixup: beyond empirical risk minimization. In: 6th International Conference on Learning Representations, ICLR 2018, Conference Track Proceedings, Vancouver, BC, Canada, 30 April–3 May 2018 (2018)
31. Zhao, F., et al.: Spherical deformable U-Net: application to cortical surface parcellation and development prediction. IEEE Trans. Med. Imaging **40**, 1217–1228 (2021)
32. Zhao, F., et al.: Spherical U-Net on cortical surfaces: methods and applications. In: IPMI (2019)

Brain Age Prediction Based on Head Computed Tomography Segmentation

Artur Paulo[1]([✉])[iD], Fabiano Filho[1][iD], Tayran Olegário[1][iD], Bruna Pinto[1][iD], Rafael Loureiro[1][iD], Guilherme Ribeiro[1][iD], Camila Silva[1][iD], Regiane Carvalho[1][iD], Paulo Santos[1][iD], Eduardo Reis[1][iD], Giovanna Mendes[1][iD], Joselisa de Paiva[1][iD], Márcio Reis[1,2][iD], and Letícia Rittner[1,3][iD]

[1] Imaging Department, Hospital Israelita Albert Einstein (HIAE), São Paulo, Brazil
`artur.marques@einstein.br`
[2] Studies and Researches in Science and Technology Group (GCITE), Instituto Federal de Goiás (IFG), Goiás, Brazil
[3] School of Electrical and Computer Engineering, Universidade de Campinas (UNICAMP), São Paulo, Brazil

Abstract. Accurate estimation of an individual's brain age holds significant potential in understanding brain development, aging, and neurological disorders. Despite the widespread availability of head computed tomography (CT) images in clinical settings, limited research has been dedicated to predicting brain age within this modality, often constrained to narrow age ranges or substantial disparities between predicted and chronological age. To address this gap, our work introduces a novel machine learning-based approach for predicting brain age using interpretable features derived from head CT segmentation. By compiling an extensive input set of characteristics including gray matter volume, white matter, cerebrospinal fluid, bone, and soft tissue, we were able to test several linear and non-linear models. Across the entire dataset, our model achieved a mean absolute error (MAE) of 6.70 years in predicting brain age. Remarkably, the relationship between bone and gray matter, as well as the volume of cerebrospinal fluid, were identified as the most pivotal features for precise brain age estimation. To summarize, our proposed methodology exhibits encouraging potential for predicting brain age using head CT scans and offers a pathway to increasing the interpretability of brain age prediction models. Future research should focus on refining and expanding this methodology to improve its clinical application and extend its impact on our understanding of brain aging and related disorders.

This work was supported by the Program of Support for the Institutional Development of the Unified Health System (PROADI-SUS,01/2020; NUP: 25000.161106/2020-61) and Albert Einstein Israelite Hospital.

Supplementary Information The online version contains supplementary material available at https://doi.org/10.1007/978-3-031-44858-4_11.

Keywords: Brain age · head computed tomography · machine
learning · aging

1   Introduction

In recent years, many studies have focused on estimating brain age based on
structural neuroimaging, [1,2], with numerous potential clinical applications,
such as early detection of neurological abnormalities, monitoring treatments,
forecasting accelerated brain atrophy, and estimating mortality risk [3–5].

Despite the dominance of Magnetic Resonance Imaging (MRI) in these studies, its cost, time consumption, and lack of ready availability in many clinical environments present significant drawbacks. Alternatively, head computed
tomography (CT) scans, more commonly performed, less costly, and rich in
valuable structural information about the brain, pose an attractive avenue for
research [6,7]. Precise estimation of brain age through head CT scans carries
substantial implications for myriad clinical applications. Moving away from traditional methods that rely on the costly and time-consuming MRI imaging, head
CT scans provide a promising platform for predicting brain age gap. Even though
it offers lower spatial resolution than MRI, head CT's accessibility and affordability make it a viable tool for neurological disorder screening [8]. The premise of
this research is that an accurate brain age estimation based on CT scans can give
clinicians and researchers valuable insights into the aging process, enabling the
identification of at-risk individuals and facilitating the creation of personalized
interventions to preserve brain health and prevent neurodegenerative disorders.

However, only a handful of studies have utilized head CT images for predicting brain age. One such study employed intracranial regions from CT images
as inputs in a convolutional neural network (CNN) to train a model predicting
brain age in pediatric brain CT images, achieving a root-mean-square-error of
7.80 months [9]. The model's training data solely consisted of newborns aged 0
to 3 years old, limiting the model's generalizability to the broader population
[9]. Another study used CT of the eye sockets (orbit) features from head CT
alongside deep convolutional ones. This study showed a median absolute error
of 9.99 years when combining images with a volumetric estimate, compared to
11.02 years with image-derived features alone and 13.28 years with volumetric
estimate features alone [10]. Despite employing a population with a broad age
range, (1–97 years old), the method only accounted for intensity and volumetric
orbital features, neglecting brain-associated changes in the CT data.

While brain age prediction based on head CT scans is a promising field
of research due to the data availability, studies using a wide age range and
brain-derived features are needed to establish its clinical validity and reliability. Furthermore, to make the models clinically applicable, they must provide
meaningful information about the brain. The widespread use of CNNs to analyze entire brain intensity images often overlooks crucial information necessary
for brain age prediction, leading to the 'black-box' problem [5]. Similarly, the
use of machine learning-derived 'weight maps' can be complex and lacks a clear

interpretation in the context of brain aging [11]. To make brain age prediction clinically applicable, it is vital to incorporate interpretable feature extraction methods, providing a more transparent and understandable approach to brain age prediction, thus enhancing its practical utility in healthcare settings.

Utilizing brain region volumes as features for brain age estimation can provide critical information about the morphometric patterns within different brain regions. With various computer algorithms and software tools assisting with brain region segmentation, the process has become quicker and more accurate than traditional manual segmentation methods [12]. Automatic brain region segmentation from head CT images via CTseg proves to be as reliable as manual segmentation and can provide pertinent information to detect brain atrophy [13], therefore showing potential for use in brain-age prediction models.

This study aims to utilize brain region segmentation from head CT images to predict brain age across a broad age range for individuals of both sexes. The specific objectives are to: i) use volumetric data obtained from white matter (WM), gray matter (GM), cerebrospinal fluid (CSF), bone, and soft tissue as features for brain age estimation, ii) explore which volumetric features are most relevant for brain age prediction, and iii) test whether there are sex differences in model fitness and performance. This study's novelty lies in its use of clinically relevant features to estimate brain age in head CT images, thus providing interpretative outcomes regarding structural changes in aging.

2 Methods

2.1 Data

We retrospectively collected a total of 694 volumetric CT scans, which exhibited no radiological findings, from Hospital Israelita Albert Einstein, a prominent tertiary hospital in São Paulo, Brazil. The dataset was compiled by searching the institutional PACS for standard CT exams, wherein the phrase 'Brain CT within normal limits' was identified within the reports of all head CT scans conducted between 2016 and 2022. The need for individual consent was waived by the Ethical Committee approval number: 58878622.0.0000.0071. Prior to analysis, DICOM metadata were anonymized, with only sex and age information retained.

2.2 Image Processing

We used a Docker Image of the CTseg tool from the Statistical Parametric Mapping (SPM) toolbox for brain segmentation. The SPM-CTseg tool employs a method defined by a joint probability distribution represented by a tissue template [14]. We segmented 693 anonymized head CT exams into GM, WM, CSF, bone and facial soft tissue (Fig. 1) using CTseg's default parameters. The masks were binarized with a threshold of 0.5 and we extracted brain volume in cm^3 using the SimpleITK library [15]. We calculated the intracranial volume (ICV) using the sum of GM volume, WM, and CSF. ICV was used to normalize the variables used in our model by dividing each variable by its value. The ratio between GM, WM and CSF was calculated and incorporated as features in the model. The proposed model's 13 features were GM volume, WM volume, CSF volume, bone volume, soft tissue volume, GM and WM ratio, soft tissue and GM ratio, CSF and GM ratio, bone and GM ratio, bone and WM ratio, estimated intracranial volume, and soft tissue and WM ratio. Each subject had their 13 features concatenated vertically.

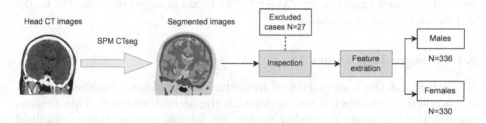

Fig. 1. Overview of the proposed method for brain age prediction: after automated segmentation by CTSeg, followed by a visual inspection of each head CT, 13 features were extracted and used to train separate regression models for males and females. In the segmented images, the soft tissue is represented by the yellow mask, bone in dark blue, cerebrospinal fluid in light blue, gray matter in green, and white matter in red. (Color figure online)

Before volume extraction, all images were visually inspected. This process involved viewing the median slice of each image in an axial view to identify any segmentation errors. Twenty-seven participants were excluded due to segmentation failures.

2.3   Age Prediction by Regression

For conducting the brain-age prediction model, subjects' ages (in years) were rounded to the nearest whole number. We leveraged the PyCaret framework [16], an open-source machine learning library that automates the machine learning workflow, to discern the optimal pipeline for predicting brain age among male and female subjects. This involved evaluating over 25 linear and non-linear regression models, as follows: Linear Regression, Lasso Regression, Ridge Regression, Elastic Net, Least Angle Regression, Lasso Least Angle Regression, Orthogonal Matching Pursuit, Bayesian Ridge, Automatic Relevance Determination, Passive Aggressive Regressor, Random Sample Consensus, TheilSen Regressor, Huber Regressor, Kernel Ridge, Support Vector Regression, K Neighbors Regressor, Decision Tree Regressor, Random Forest Regressor, Extra Trees Regressor, AdaBoost Regressor, Gradient Boosting Regressor MLP Regressor, Extreme Gradient Boosting, Light Gradient Boosting MachineDummy Regressor. The efficiency of the pipeline was gauged based on minimizing the mean absolute error (MAE) between the predicted and actual chronological age of the subjects.

PyCaret streamlined the process and provided us with effective tools for model selection and evaluation. To do this, we randomly selected 70% of the subjects from the dataset. We generated two samples stratifying by sex, resulting in 230 female and 230 male subjects for the training set. For the test dataset, 100 female and 100 male subjects were used. The same number of samples were used for the model concerning both sexes (chosen randomly). Features demonstrating high multicollinearity, with a Pearson correlation above 0.95, were subsequently removed. Model performance was evaluated using 10-fold cross-validation. The optimal pipelines suggested by PyCaret were then employed to create the model and extract feature importance.

2.4   Feature Importance

To understand the contribution of individual morphological features to brain age prediction, we selected the model with the highest accuracy. This analysis contemplated separate regression models for females, males, and a combined model for both sexes. Using the kernel Shapley additive explanation (SHAP) method [17], we investigated the regional morphological features contributing to the model prediction error for each of the three best-performing algorithms.

3   Results

3.1   Demographic Information

Our data set comprised individuals ranging from 0 to 84 years old, with a mean age of 23.60 (SD = 13.98) years. The mean age for females was 23.36 (SD = 12.96), and for males, it was 23.85 (SD = 12.97). There was no significant difference in terms of age between males and females (p = 0.65), as illustrated in Fig. 2(a).

3.2   Model Performance

The features with high collinearity, which were excluded from all models, included: Bone and WM ratio, Bone volume, CSF and GM ratio, Soft tissue volume, Soft tissue and WM ratio (see Supplementary Materials S1). The Extra Tree Regressor model demonstrated the highest performance for the female dataset and the Linear Regressor was the best for the male dataset. The Extra Tree Regressor showed the best performance when considering both sexes. Please refer to Tables S2 in the Supplementary Materials for a detailed description of the results from all tested models in the female and male datasets. Table 1 presents the top five best models for females, males, and both sexes.

   We validated the best models using 10-fold cross validation. For the model concerning females, the mean MEA value was 6.827 (SD = 0.972) and R^2 of 0.585 (SD = 0.119). For males, the mean MEA value was 7.571 (SD = 1.35) and R^2 of 0.39 (SD = 0.147). For the model concerning both sexes, the mean MAE was

Table 1. Results obtained for the top five models that showed the best performance. MAE is mean absolute error, RMSE is root mean square error and R^2 is coefficient of determination

Sample	Model	MAE	RMSE	R^2
Females	**Extra Trees Regressor**	6.7987	8.8751	0.5938
	Gradient Boosting Regressor	6.875	8.8977	0.5898
	Random Forest Regressor	6.9225	9.072	0.576
	Light Gradient Boosting Machine	6.9228	9.0179	0.5807
Males	**Linear Regression**	7.5713	9.4638	0.394
	Least Angle Regression	7.5733	9.4255	0.3968
	Automatic Relevance Determination	7.6333	9.5082	0.3882
	Bayesian Ridge	7.6578	9.5354	0.3913
All subjects	**Extra Trees Regressor**	6.6142	8.7026	0.5355
	Random Forest Regressor	6.9957	9.3149	0.4729
	AdaBoost Regressor	7.2228	9.151	0.4948
	Light Gradient Boosting Machine	7.2788	9.6004	0.4342

6.702 (SD = 1.530). We used an independent sample t-test to test whether the model's MAE and R^2 were different for each sex. A difference was detected for R^2 (t = −2.522, p = 0.0189), but not for MAE (t = 0.845, p = 0.406). Figure 2(b) displays the real versus predicted age by the model for each sex. We calculated correlation coefficients between the real and predicted age using the Spearman method. The correlation was strong for females (r = 0.831, p < 0.0001) and moderate for males (r = 0.657, p < 0.0001).

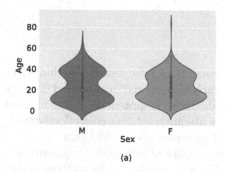

(a)

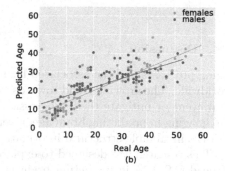

(b)

Fig. 2. (a) Age distribution between males (M) and females (F). (b) Chronological age vs. predicted age using best models for each (separated by sex).

3.3   Feature Importance

We assessed the feature importance for each of the groups using Variable explicability provided by Shapley Additive exPlanations (SHAP). For females, the most important feature was bone volume and GM ratio, CSF, followed by soft tissue and GM ratio. For the male group, the most important feature was CSF volume, followed by WM, GM volume, and bone and GM ratio.

For the whole dataset, including sex as a feature, results indicate that the bone and GM ratio and the CSF volume are the top two most important features, as shown in Fig. 3.

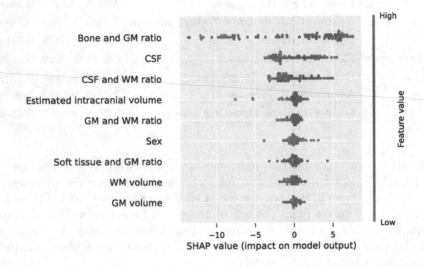

Fig. 3. Variable explicability provided by Shapley Additive exPlanations (SHAP).

3.4   Discussion

Aging processes and the associated health conditions pose significant challenges worldwide. The growing trend towards the early detection of age-related diseases to prevent or mitigate their progression highlights the potential utility of head CT imaging in predicting brain age, given its ubiquity, relative cost-effectiveness, and the invaluable structural information it provides about the brain.

This research was designed to explore the utility of automated segmentation of head CT data in predicting brain age. We assessed three distinct predictive models using samples from males, females, and a combination of both sexes. We executed volume segmentation of brain regions, incorporating gray matter (GM), white matter (WM), soft tissue, bone, and cerebrospinal fluid (CSF), in order to extract the features for our models.

Our study achieved a mean absolute error (MAE) of 6.70 years for the entire dataset (i.e., both sexes), an outcome that is commensurate with prior studies employing analogous methods. Previous research using MRI and comparable machine learning methods, specifically region of interest analysis, have reported MAE values ranging from 4.5 to 6.59 years in healthy individuals [18–21].

Our model displayed superior performance when compared with earlier studies that employed head CT data for brain age prediction. One such study, which utilized orbital features derived from head CT scans alongside volumetric features, reported a median absolute error of 9.99 years [10]. Another study employing convolutional neural networks (CNN) in pediatric CT yielded a root-mean-square error of 7.8 months, representing 21.66% of the total age range within their dataset. In comparison, our MAE value corresponds to 8.9%. It is crucial, however, to consider the disparate populations utilized to test these models and exercise caution when interpreting variations in model performance. Nevertheless, our study illustrates that even without the use of CNN, but through a focus on pertinent volumetric features of the brain parenchyma, we could obtain comparable and meaningful data for brain age prediction.

In our model pertaining to both sexes, the bone to GM ratio and CSF volume emerged as critical features. SHAP values imply that larger values of bone volume and its ratio with gray matter correlate with higher predictive values for brain age. Prior research has demonstrated that cranial bones thicken [22] and undergo shape changes [23] as individuals age. This phenomenon could be ascribed to the adaptive mechanism of skull thickening, which serves to maintain intracranial pressure in response to the decrease in brain parenchyma volume associated with aging [22,24]. These findings suggest that alterations in bone volume and its ratio with gray matter significantly contribute to the prediction of brain age and potentially mirror age-related adaptations within the skull. The previously mentioned study, which employed convolutional neural networks (CNN) to predict brain age based on orbital features, underscores the significant contribution of skull bones as robust indicators of aging, particularly in older individuals, as evidenced through activation maps [10].

Conversely, increased CSF values also correlate with higher brain age. This correlation might be attributed to the expansion of ventricles and sulcal spaces typically observed as individuals age [25]. The enlarged CSF volume, possibly due to these age-related structural changes in the brain, is manifested in higher brain age estimates.

The comparison of R^2 reveals a difference between sexes but no significant difference in MAE, implying that females have a better model fit to the data, which, however, does not impact the model's overall performance. Differences in brain age predictive models between males and females can be attributed to various factors, such as sex-based differences in brain structure, cognitive factors, and behavioral factors. Notably, changes in skull shape and intracranial volume are influenced by sex [23,24]. Given that our sample was evenly distributed between sexes, alterations in the model's goodness-of-fit for age prediction could be ascribed to distinct patterns of skull and brain parenchyma distribution between sexes during aging.

Estimating brain age at a localized level within the brain could yield spatial information on anatomical patterns of brain aging [26], allowing for differential age estimation in specific brain tissues, and thereby providing valuable data for future clinical use. Prior work has revealed group differences in mild cognitive impairment or dementia, particularly in subcortical regions like the accumbens, putamen, pallidum, hippocampus, and amygdala [26].

It is important to underscore that this study did not employ convolutional neural networks (CNNs) for the analysis. This approach offers noteworthy benefits. Firstly, the lack of dependency on CNNs allowed for quicker training of the model. As CNNs generally require a significant amount of computational power and time for training, our model offers a more efficient and expeditious alternative. Secondly, a reduced reliance on computational resources makes our approach more accessible for research settings that may have limited computational capacities.

The present study has certain limitations. Our estimation of the MAE ratio, considering the age range, could potentially misrepresent the actual situation due to the pronounced variability in the age distribution of our sample. Additionally, the absence of data regarding cognitive abilities and the potential undetected subtle neurodegenerative processes during radiological evaluations warrant further consideration. The integration of neuropsychological data in future work could provide a more holistic understanding of brain aging. Subsequent research should focus on refining brain age prediction in head CTs, perhaps by including T1-MRI data, and constructing models that can predict distinct brain tissue ages for more clinically pertinent outcomes.

4   Conclusion

The objective of our study was to develop a model for predicting brain age by employing automatic segmentation of head CT scans. We illustrate that accurate brain age prediction can be achieved by extracting the volume of brain region segmentation from head CT scans. While there are differences in model performance between sexes, stratifying the sample by sex does not yield improved outcomes. Notably, the most influential feature for predicting brain age is the relationship between bone and gray matter. Remarkably, the volumetric features obtained from automatic segmentation of CT scans display comparable performance to contemporary CNN models. As we move forward, it is imperative for future research to focus on developing models capable of predicting distinct brain tissue ages, thereby yielding more interpretable and clinically relevant results.

References

1. Mishra, S., Beheshti, I., Khanna, P.: A review of neuroimaging-driven brain age estimation for identification of brain disorders and health conditions. IEEE Rev. Biomed. Eng. **16**, 371–385 (2021)

2. Baecker, L., Garcia-Dias, R., Vieira, S., Scarpazza, C., Mechelli, A.: Machine learning for brain age prediction: introduction to methods and clinical applications. EBioMedicine **72**, 103600 (2021)
3. Cole, J.H., Leech, R., Sharp, D.J., Alzheimer's Disease Neuroimaging: Initiative Prediction of brain age suggests accelerated atrophy after traumatic brain injury. Ann. Neurol. **77**(4), 571–581 (2015)
4. Cole, J.H., et al.: Brain age predicts mortality. Mol. Psychiatry **23**(5), 1385–1392 (2018)
5. Cole, J.H., Franke, K.: Predicting age using neuroimaging: innovative brain ageing biomarkers. Trends Neurosci. **40**(12), 681–690 (2017)
6. Wippold, F.J.: Head and neck imaging: the role of CT and MRI. J. Magn. Reson. Imaging Official J. Int. Soc. Magn. Reson. Med. **25**(3), 453–465 (2007)
7. Vymazal, J., Rulseh, A.M., Keller, J., Janouskova, L.: Comparison of CT and MR imaging in ischemic stroke. Insights Imaging **3**(6), 619–627 (2012)
8. McLane, H.C., et al.: Availability, accessibility, and affordability of neurodiagnostic tests in 37 countries. Neurology **85**(18), 1614–1622 (2015)
9. Morita, R., et al.: Brain development age prediction using convolutional neural network on pediatrics brain CT images, pp. 1–6 (2021)
10. Bermudez, C., et al.: Anatomical context improves deep learning on the brain age estimation task. Magn. Reson. Imaging **62**, 70–77 (2019)
11. Haufe, S., et al.: On the interpretation of weight vectors of linear models in multivariate neuroimaging. Neuroimage **87**, 96–110 (2014)
12. Mahender Kumar Singh and Krishna Kumar Singh: A review of publicly available automatic brain segmentation methodologies, machine learning models, recent advancements, and their comparison. Ann. Neurosci. **28**(1–2), 82–93 (2021)
13. Adduru, V., et al.: A method to estimate brain volume from head CT images and application to detect brain atrophy in Alzheimer disease. Am. J. Neuroradiol. **41**(2), 224–230 (2020)
14. Brudfors, M., Balbastre, Y., Flandin, G., Nachev, P., Ashburner, J.: Flexible Bayesian modelling for nonlinear image registration. In: Martel, A.L., et al. (eds.) MICCAI 2020. LNCS, vol. 12263, pp. 253–263. Springer, Cham (2020). https://doi.org/10.1007/978-3-030-59716-0_25
15. Lowekamp, B.C., Chen, D.T., Ibanez, L., Blezek, D.: The design of SimpleITK. Front. Neuroinform. **45**(7), 1–14 (2013)
16. Ali, M.: PyCaret: an open source, low-code machine learning library in Python. PyCaret version, 2 (2020)
17. Lundberg, S.M., Lee, S.-I.: A unified approach to interpreting model predictions. In: Advances in Neural Information Processing Systems, vol. 30 (2017)
18. Han, L.K.M.: Brain aging in major depressive disorder: results from the enigma major depressive disorder working group. Mol. Psychiatry **26**, 5124–5139 (2021)
19. Ly, M.: Improving brain age prediction models: incorporation of amyloid status in Alzheimer's disease. Neurobiol. Aging **87**, 44–48 (2020)
20. Franke, K., Gaser, C., Manor, B., Novak, V.: Advanced BrainAGE in older adults with type 2 diabetes mellitus. Front. Aging Neurosci. **5**, 90 (2013)
21. Lancaster, J., Lorenz, R., Leech, R., Cole, J.H.: Bayesian optimization for neuroimaging pre-processing in brain age classification and prediction. Front. Aging Neurosci. **10**(28), 1–10 (2018)
22. May, H., Mali, Y., Dar, G., Abbas, J., Hershkovitz, I., Peled, N.: Intracranial volume, cranial thickness, and hyperostosis frontalis interna in the elderly. Am. J. Hum. Biol. **24**(6), 812–819 (2012)

23. Urban, J.E., Weaver, A.A., Lillie, E.M., Maldjian, J.A., Whitlow, C.T., Stitzel, J.D.: Evaluation of morphological changes in the adult skull with age and sex. J. Anat. **229**, 838–846 (2016)
24. Royle, N.A., et al.: Influence of thickening of the inner skull table on intracranial volume measurement in older people. Magn. Reson. Imaging **31**(6), 918–922 (2013)
25. Longstreth, W.T., Jr., et al.: Clinical correlates of ventricular and sulcal size on cranial magnetic resonance imaging of 3,301 elderly people: the cardiovascular health study. Neuroepidemiology **19**(1), 30–42 (2000)
26. Popescu, S.G., Glocker, B., Sharp, D.J., Cole, J.H.: Local brain-age: a U-Net model. Front. Aging Neurosci. **13**, 761954 (2021)

Pretraining is All You Need: A Multi-Atlas Enhanced Transformer Framework for Autism Spectrum Disorder Classification

Lucas Mahler[1(✉)], Qi Wang[1], Julius Steiglechner[1,2], Florian Birk[1,2], Samuel Heczko[1], Klaus Scheffler[1,2], and Gabriele Lohmann[1,2]

[1] Max-Planck-Institute for Biological Cybernetics, 72076 Tübingen, Germany
lucas.mahler@tuebingen.mpg.de
[2] University Hospital Tübingen, 72076 Tübingen, Germany

Abstract. Autism spectrum disorder (ASD) is a prevalent psychiatric condition characterized by atypical cognitive, emotional, and social patterns. Timely and accurate diagnosis is crucial for effective interventions and improved outcomes in individuals with ASD. In this study, we propose a novel Multi-Atlas Enhanced Transformer framework, METAFormer, ASD classification. Our framework utilizes resting-state functional magnetic resonance imaging data from the ABIDE I dataset, comprising 406 ASD and 476 typical control (TC) subjects. METAFormer employs a multi-atlas approach, where flattened connectivity matrices from the AAL, CC200, and DOS160 atlases serve as input to the transformer encoder. Notably, we demonstrate that self-supervised pretraining, involving the reconstruction of masked values from the input, significantly enhances classification performance without the need for additional or separate training data. Through stratified cross-validation, we evaluate the proposed framework and show that it surpasses state-of-the-art performance on the ABIDE I dataset, with an average accuracy of 83.7% and an AUC-score of 0.832. The code for our framework is available at github.com/Lugges991/METAFormer.

Keywords: Deep Learning · Transformers · fMRI · Autism Spectrum Disorder Classification

1 Introduction

Autism spectrum disorder (ASD) is a widespread psychiatric condition characterized by atypical cognitive, emotional, and social patterns. With millions of individuals affected worldwide, the early diagnosis of ASD is a critical research priority, as it has a significant positive impact on patient outcomes. The etiology of ASD remains elusive, with intricate interactions among genetic, biological, psychological, and environmental factors playing a role. Currently, diagnosing ASD relies heavily on behavioral observations and anamnestic information, posing challenges and consuming a considerable amount of time. Skilled clinicians with extensive experience are required for accurate diagnosis. However,

© The Author(s), under exclusive license to Springer Nature Switzerland AG 2023
A. Abdulkadir et al. (Eds.): MLCN 2023, LNCS 14312, pp. 123–132, 2023.
https://doi.org/10.1007/978-3-031-44858-4_12

common assessments of ASD have been criticized for their lack of objectivity and transparency [27]. Given these limitations, there is an urgent need for a fast, cost-effective, and objective diagnostic method that can accurately identify ASD leading to more timely interventions and improved outcomes for affected individuals.

In recent years, magnetic resonance imaging (MRI) has emerged as a powerful non-invasive tool for gaining insights into brain disorders' pathophysiology. Functional MRI (fMRI), a notable advancement in MRI technology, allows for the investigation of brain function by measuring changes in blood oxygen levels over time. Functional connectivity (FC) analysis [4] plays a crucial role in fMRI data analysis, as it examines the statistical dependencies and temporal correlations among different brain regions. Rather than considering isolated abnormalities in specific regions, brain disorders often arise from disrupted communication and abnormal interactions between regions. FC analysis enables researchers to explore network-level abnormalities associated with various disorders. This analysis involves partitioning the brain into regions of interest (ROIs) and quantifying the correlations between their time series using various mathematical measures.

In recent years machine learning approaches have been widely applied to the problem of ASD classification using resting-state fMRI (rs-fMRI) data. The majority of these studies use functional connectivities obtained from a predefined atlas as input to their classifiers. A considerable amount of work used classical machine learning algorithms such as support vector machines and logistic regression to classify ASD [11]. However, these methods have limitations as they are typically applied to small datasets with specific protocols and fixed scanner parameters, which may not adequately capture the heterogeneity present in clinical data. 3D Convolutional neural networks [20,26,33] have also been applied to preprocessed fMRI data, [1] have used 2D CNNs on preprocessed fMRI data. Though, these approaches are as well limited by the fact that they were only using small homogeneous datasets.

More recent works tried to overcome the homogeneity limitations and have used deep learning approaches to classify ASD based on connectomes. Multilayer perceptrons are suited to the vector based representations of connectomes and have thus seen some usage in ASD classification [12,30]. Graph convolutional models are also well suited and have yielded high accuracies [19,29]. Other approaches used 1D CNNs [23], or variants of recurrent neural networks [17,18], and also probabilistic neural networks have been proposed [16].

However, ASD classification is not limited to fMRI data and there has been work using, for example, EEG [5] or also more novel imaging approaches such as functional near-infrared spectroscopy [13].

The current study aims to improve classification performance of ASD based on rs-fMRI data over the entire ABIDE I dataset [22] by leveraging the representational capabilities of modern transformer architectures. We thus summarize our main contributions as follows:

1. We propose a novel multi-atlas enhanced transformer framework for ASD classification using rs-fMRI data: METAFormer

2. We demonstrate that self-supervised pretraining leads to significant improvements in performance without the requirement of additional data.
3. We show that our model outperforms state of the art methods on the ABIDE I dataset.

2   Methods

2.1   Dataset

Our experiments are conducted on the ABIDE I dataset [22] which is a publicly available dataset containing structural MRI as well as rs-fMRI data obtained from individuals with Autism Spectrum Disorder (ASD) and typical controls (TC) from 17 different research sites. The raw dataset encompasses a total of 1112 subjects, 539 of which are diagnosed with ASD and 573 are TC. Subjects ages range from 7 to 64 years with a median age of 14.7 years across groups. The ABIDE I dataset is regarded as one of the most comprehensive and widely used datasets in the field, offering a combination of MRI, rs-fMRI, and demographic data.

The ABIDE I dataset exhibits significant heterogeneity and variations that should be taken into account. It comprises data from diverse research sites worldwide, leading to variations in scanning protocols, age groups, and other relevant factors. Consequently, the analysis and interpretation of the ABIDE I dataset pose challenges due this inherent heterogeneity.

Preprocessing Pipeline. We utilize the ABIDE I dataset provided by the Preprocessed Connectomes Project (PCP) [6] for our analysis. The PCP provides data for ABIDE I using different preprocessing strategies. In this work we use the preprocessed data from the DPARSF pipeline [31] comprising 406 ASD and 476 TC subjects. The DPARSF pipeline is based on SPM8 and includes the following steps: The first 4 volumes of each fMRI time series are discarded to allow for magnetization stabilization. Slice timing correction is performed to correct for differences in acquisition time between slices. The fMRI time series are then realigned to the first volume to correct for head motion. Intensity normalization is not performed. To clean confounding variation due to physiological noise, 24-parameter head motion, mean white matter and CSF signals are regressed out. Motion realignment parameters are also regressed out as well as linear and quadratic trends in low-frequency drifts. Bandpass filtering was performed after regressing nuisance signals to remove high-frequency noise and low-frequency drifts. Finally, functional to anatomical registration is performed using rigid body transformation and anatomical to standard space registration is performed using DARTEL [2].

Functional Connectivity. As the dimensionality of the preprocessed data is very high, we perform dimensionality reduction by dividing the brain into a set of parcels or regions with similar properties according to a brain atlas. In this work we process our data using three different atlases. The first atlas is the Automated Anatomical Labeling (AAL) atlas [25]. This atlas, which is widely used in the literature, divides the brain into 116 regions of interest (ROIs) based on anatomical landmarks and was fractionated to functional resolution of $3\,mm^3$ using nearest-neighbor interpolation. The second atlas is the Craddock 200 (CC200) atlas [7]. It divides the brain into 200 ROIs based on functional connectivity and was fractionated to functional resolution of $3\,mm^3$ using nearest-neighbor interpolation. The third atlas we considered is the Dosenbach 160 (DOS160) atlas [10] which contains uniform spheres placed at coordinates obtained from meta-analyses of task-related fMRI studies.

After obtaining the ROI time-series from the atlases, we compute the functional connectivity using the Pearson Correlation Coefficient between each pair of ROIs. The upper triangular part of the correlation matrix as well as the diagonal are then dropped and the lower triangular part is vectorized to obtain a feature vector of length $k(k-1)/2$, where k is the number of ROIs, which is then standardized and serves as input to our models.

2.2   Model Architecture

METAFormer: Multi-Atlas Transformer. Here, we propose METAFormer, at the core of which lies the transformer encoder architecture, originally proposed by Vaswani et al. [28] for natural language processing tasks. However, as our main goal is to perform classification and not generation we do not use the decoder part of the transformer architecture. In order to accommodate input from multiple different atlases, we employ an ensemble of three separate transformers, with each transformer corresponding to a specific atlas. As depicted in Fig. 1, the input to each transformer is a batch of flattened functional connectivity matrices. First, the input to each transformer undergoes embedding into a latent space using a linear layer with a dimensionality of $d_{model} = 256$. The output of the embedding is then multiplied by $\sqrt{d_{model}}$ to scale the input features. This scaling operation aids in balancing the impact of the input features with the attention mechanism. Since we are not dealing with sequential data, positional encodings are not utilized.

The embedded input is subsequently passed through a BERT-style encoder [9], which consists of $N = 2$ identical layers with $d_{ff} = 128$ feed forward units, and $h = 4$ attention heads. To maintain stability during training, each encoder layer is normalized using layer normalization [3], and GELU [15] is used as the activation function. Following the final encoder layer, the output passes through a dropout layer. Then, a linear layer with d_{model} hidden units and two output units corresponding to the two classes is applied to obtain the final output. The outputs of the three separate transformers are averaged, and this averaged representation is passed through a softmax layer to derive the final class probabilities.

To train our Multi-Atlas Transformer model, we follow a series of steps. Firstly, we initialize the model weights using the initialization strategy proposed by He [14], while setting the biases to 0. To optimize the model, we employ the

AdamW optimizer [21] and minimize the binary cross entropy between predictions and labels. Our training process consists of 750 epochs, utilizing a batch size of 256. To prevent overfitting, we implement early stopping with a patience of 40 epochs. In order to ensure robustness of our model, we apply data augmentation. Specifically, we randomly introduce noise to each flattened connectome vector with an augmentation probability of 0.3. The noise standard deviation is set to 0.01. We conduct hyperparameter tuning using grid search. We optimize hyperparameters related to the optimizer, such as learning rate and weight decay. We also consider the dropout rate during this process.

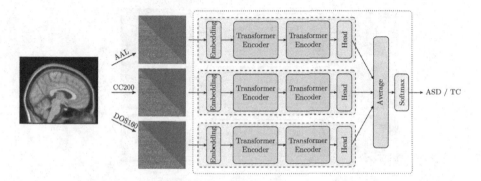

Fig. 1. Architecture of METAFormer. Our model consists of three separate transformers, each corresponding to a specific atlas. The input to each transformer is a batch of flattened functional connectivity matrices with the diagonal and the upper triangular part of the matrix removed. The output of the transformers is averaged and passed through a softmax layer to derive the final class probabilities.

2.3   Self-supervised Pretraining

As popularized by [24], the utilization of self-supervised generative pretraining followed by task-specific fine-tuning has demonstrated improved performance in transformer architectures. Building upon this approach, we propose a self-supervised pretraining task for our model. Our approach involves imputing missing elements in the functional connectivity matrices, drawing inspiration from the work introduced by [32]. To simulate missing data, we randomly set 10% of the standardized features in each connectome to 0 and train the model to predict the missing values. The corresponding configuration is illustrated in Fig. 2. To achieve this, we begin by randomly sampling a binary noise mask $M \in \{0,1\}^{n_i}$ for each training sample, where n_i denotes the number of features in the i-th connectome. Subsequently, the original input X is masked using element-wise multiplication with the noise mask: $X_{masked} = X \odot M$.

To estimate the corrupted input, we introduce a linear layer with n_i output neurons on top of the encoder stack, which predicts $\hat{x}_i$. We calculate a multi atlas masked mean squared error (MAMSE) loss $\mathcal{L}_{multi}$ between the predicted and the original input:

$$\mathcal{L}_{multi} = \frac{1}{3} \sum_{i=1}^{3} \frac{1}{n_i} \sum_{j \in M}^{n_i} ||x_{i,j} - \hat{x}_{i,j}||^2 \tag{1}$$

where $x_{i,j}$ is the original value of the j-th masked input from the i-th atlas and $\hat{x}_{i,j}$ is the predicted value for the masked input at position j in the i-th atlas.

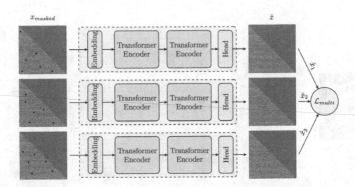

Fig. 2. Self-supervised pretraining on the imputation task of the METAFormer architecture. The inputs to the model are masked connectomes, where 10% of the features are randomly set to 0 (exemplified as black squares). The model is trained to predict the missing values implying that the output of the model has the same shape as the input.

3 Experiments

3.1 Experimental Setup

To evaluate the classification performance of our models in a robust manner, we employed 10-fold stratified cross-validation. For each fold, the model is trained on the remaining nine training folds and evaluated on the held-out test fold. Further, we set aside 30% of each training fold as validation sets which are then used for hyperparameter tuning and early stopping.

In order to assess the impact of self-supervised pretraining, we compare the performance of our model with and without pretraining. To achieve that, we first pretrain the model using the imputation task on the training folds and subsequently fine-tune the model on the training folds using the classification task after which we evaluate on the held-out test fold.

In order to verify the efficacy of using multiple atlases as input we compared the performance of our METAFormer model with the performance of single atlas transformer (SAT) models. For that, we trained three separate transformer models using only one atlas as input. The SAT models are trained using the same architecture as well as training procedure as the METAFormer model. We also evaluated the performance of the SAT models with and without self-supervised pretraining in order to asses its impact on the performance of the model. To make results comparable, we use the same training and validation folds for all model configurations under investigation.

3.2   Evaluation Metrics

By using cross-validation, we obtained 10 different sets of performance scores per configuration. These scores were then averaged and the standard deviation of each score was obtained, providing reliable estimates of the model's performance on unseen data. The classification results were reported in terms of accuracy, precision, recall (sensitivity), F1-score and AUC-score, which are commonly used metrics for evaluating classification models.

4   ASD Classification Results

Table 1 shows the superior performance of our pretrained METAFormer model compared to previously published ASD classifiers trained on atlas-based connectomes. Importantly, our model achieves higher accuracy even when compared to approaches with similar test set sizes that did not employ cross-validation.

To further validate the effectiveness of our proposed Multi-Atlas Transformer model for Autism Spectrum Disorder classification, we compare METAFormer against single atlas transformers. The results, as presented in Table 2, demonstrate the superiority of METAFormer over all single atlas models in terms of accuracy, precision, recall, F1-score, and AUC-score. Moreover, the multi-atlas model exhibits comparable or lower standard deviations in performance metrics compared to the single atlas models. This indicates higher robustness and stability of our multi-atlas METAFormer architecture, attributed to the joint training of the three transformer encoders.

4.1   Impact of Pretraining

We also evaluated the effect of self-supervised pretraining on the classification performance of our models. As Table 2 shows pretraining significantly improves the performance of all models in terms of accuracy, precision, recall, F1-score and AUC-score. Furthermore, for our proposed METAFormer architecture pretraining improves the performance by a large margin.

Table 1. Overview of state-of-the-art ASD classification methods that use large, heterogenous samples from ABIDE I. Note that our model achieves the highest accuracy while still using 10-fold cross-validation.

Study	#ASD	#TC	Model	CV	Acc.
MAGE [29]	419	513	Graph CNN	10-fold	75.9%
AIMAFE [30]	419	513	MLP	10-fold	74.5%
1DCNN-GRU [23]	–	–	1D CNN	–	78.0%
MISODNN [12]	506	532	MLP	10-fold	78.1%
3D CNN [8]	539	573	3D CNN	5-fold	74.53%
CNNGCN [17]	403	468	CNN/GRU	–	72.48%
SSRN [18]	505	530	LSTM	–	81.1%
Ours	408	476	Transformer	10-fold	**83.7%**

Table 2. Classification results for the different model configurations. Reported values are the mean ± standard deviation over 10 folds. The best results are in bold. SAT = Single Atlas Transformer, PT = Pretrained, atlases are in braces. Note that pretraining significantly improves performance across metrics and atlases. Using our multi-atlas METAFormer in combination with pretraining yields impressive performance increases.

Variant	Acc.	Prec.	Rec.	F1	AUC
METAFormer PT	**0.837** ±0.030	**0.819**±0.045	0.901 ±0.044	**0.856** ±0.023	**0.832** ±0.032
METAFormer	0.628 ±0.041	0.648±0.040	0.688 ±0.091	0.663 ±0.047	0.623 ±0.041
SAT (AAL)	0.593 ±0.040	0.585±0.042	0.888 ±0.091	0.701 ±0.024	0.568 ±0.047
SAT (CC200)	0.586 ±0.037	0.577±0.027	0.888 ±0.057	0.698 ±0.019	0.560 ±0.044
SAT (DOS160)	0.570 ±0.055	0.571±0.038	0.816 ±0.101	0.670 ±0.051	0.550 ±0.056
SAT (AAL) PT	0.601 ±0.069	0.587±0.055	0.939 ±0.059	0.719 ±0.033	0.573 ±0.077
SAT (CC200) PT	0.632 ±0.071	0.622±0.074	0.891 ±0.102	0.724 ±0.035	0.611 ±0.082
SAT (DOS160) PT	0.683 ±0.094	0.652±0.091	**0.964** ±0.057	0.771 ±0.047	0.660 ±0.106

5   Conclusion

In this paper, we propose METAFormer, a novel multi-atlas enhanced pretrained transformer architecture for ASD classification. We utilize self-supervised pretraining on the imputation task on the same dataset to prime the model for the downstream task. We conducted extensive experiments to demonstrate the effectiveness of our approach by comparing it to several baselines that use single-atlas and multi-atlas approaches with and without pretraing. Our results show that our model performs better than state-of-the-art methods and that pretraining is highly beneficial for the downstream task.

Acknowledgements. The authors thank the International Max Planck Research School for the Mechanisms of Mental Function and Dysfunction (IMPRS-MMFD) for supporting Samuel Heczko. Florian Birk is supported by the Deutsche Forschungsgesellschaft (DFG) Grant DFG HE 9297/1-1. Julius Steiglechner is funded by Alzheimer Forschung Initiative e.V.Grant #18052.

References

1. Ahammed, M.S., Niu, S., Ahmed, M.R., Dong, J., Gao, X., Chen, Y.: DarkASD-Net: classification of ASD on functional MRI using deep neural network. Front. Neuroinf. **15**, 635657 (2021)
2. Ashburner, J.: A fast diffeomorphic image registration algorithm. Neuroimage **38**(1), 95–113 (2007)
3. Ba, J.L., Kiros, J.R., Hinton, G.E.: Layer normalization (2016)
4. Biswal, B., Yetkin, F.Z., Haughton, V.M., Hyde, J.S.: Functional connectivity in the motor cortex of resting human brain using echo-planar MRI. Magn. Reson. Med. **34**(4), 537–541 (1995)
5. Brihadiswaran, G., Haputhanthri, D., Gunathilaka, S., Meedeniya, D., Jayarathna, S.: EEG-based processing and classification methodologies for autism spectrum disorder: a review. J. Comput. Sci. **15**(8), 1161–1183 (2019)
6. Craddock, C., et al.: The neuro bureau preprocessing initiative: open sharing of preprocessed neuroimaging data and derivatives. Front. Neuroinform. **7**, 27 (2013)
7. Craddock, R.C., James, G., Holtzheimer, P.E., III., Hu, X.P., Mayberg, H.S.: A whole brain fMRI atlas generated via spatially constrained spectral clustering. Hum. Brain Mapp. **33**(8), 1914–1928 (2012)
8. Deng, J., Hasan, M.R., Mahmud, M., Hasan, M.M., Ahmed, K.A., Hossain, M.Z.: Diagnosing autism spectrum disorder using ensemble 3D-CNN: a preliminary study. In: 2022 IEEE International Conference on Image Processing (ICIP), October 2022. IEEE (2022)
9. Devlin, J., Chang, M.W., Lee, K., Toutanova, K.: BERT: pre-training of deep bidirectional transformers for language understanding (2019)
10. Dosenbach, N.U.F., et al.: Prediction of individual brain maturity using fMRI. Science **329**(5997), 1358–1361 (2010)
11. Du, Y., Fu, Z., Calhoun, V.D.: Classification and prediction of brain disorders using functional connectivity: promising but challenging. Front. Neurosci. **12**, 1–29 (2018)
12. Epalle, T.M., Song, Y., Liu, Z., Lu, H.: Multi-atlas classification of autism spectrum disorder with hinge loss trained deep architectures: abide i results. Appl. Soft Comput. **107**, 107375 (2021)
13. Gerloff, C., Konrad, K., Kruppa, J., Schulte-Rüther, M., Reindl, V.: Autism spectrum disorder classification based on interpersonal neural synchrony: can classification be improved by dyadic neural biomarkers using unsupervised graph representation learning? In: Abdulkadir, A., et al. (eds.) Machine Learning in Clinical Neuroimaging, MLCN 2022. LNCS, vol. 13596, pp. 147–157. Springer, Cham (2022). https://doi.org/10.1007/978-3-031-17899-3_15
14. He, K., Zhang, X., Ren, S., Sun, J.: Delving deep into rectifiers: surpassing human-level performance on ImageNet classification (2015)
15. Hendrycks, D., Gimpel, K.: Gaussian error linear units (GELUs) (2023)
16. Iidaka, T.: Resting state functional magnetic resonance imaging and neural network classified autism and control. Cortex **63**, 55–67 (2015)
17. Jiang, W., et al.: CNNG: a convolutional neural networks with gated recurrent units for autism spectrum disorder classification. Front. Aging Neurosci. **14**, 948704 (2022)

18. Kang, L., Gong, Z., Huang, J., Xu, J.: Autism spectrum disorder recognition based on machine learning with ROI time-series. NeuroImage Clin. (2023)
19. Lamani, M.R., Benadit, P.J., Vaithinathan, K.: Multi-atlas graph convolutional networks and convolutional recurrent neural networks-based ensemble learning for classification of autism spectrum disorders. SN Comput. Sci. **4**(3), 213 (2023)
20. Li, X., Dvornek, N.C., Zhuang, J., Ventola, P., Duncan, J.S.: Brain biomarker interpretation in ASD using deep learning and fMRI. In: Frangi, A.F., Schnabel, J.A., Davatzikos, C., Alberola-López, C., Fichtinger, G. (eds.) MICCAI 2018. LNCS, vol. 11072, pp. 206–214. Springer, Cham (2018). https://doi.org/10.1007/978-3-030-00931-1_24
21. Loshchilov, I., Hutter, F.: Decoupled weight decay regularization (2019)
22. Martino, A.D., Yan, C.G., et al.: The autism brain imaging data exchange: towards a large-scale evaluation of the intrinsic brain architecture in autism. Mol. Psychiatry **19**(6), 659–667 (2013)
23. Qayyum, A., et al.: An efficient 1DCNN-LSTM deep learning model for assessment and classification of fMRI-based autism spectrum disorder. In: Raj, J.S., Kamel, K., Lafata, P. (eds.) Innovative Data Communication Technologies and Application, vol. 96, pp. 1039–1048. Springer, Singapore (2022). https://doi.org/10.1007/978-981-16-7167-8_77
24. Radford, A., Narasimhan, K., Salimans, T., Sutskever, I., et al.: Improving language understanding by generative pre-training (2018)
25. Rolls, E.T., Huang, C.C., Lin, C.P., Feng, J., Joliot, M.: Automated anatomical labelling atlas 3. Neuroimage **206**, 116189 (2020)
26. Thomas, R.M., Gallo, S., Cerliani, L., Zhutovsky, P., El-Gazzar, A., van Wingen, G.: Classifying autism spectrum disorder using the temporal statistics of resting-state functional MRI data with 3D convolutional neural networks. Front. Psychiatry **11**, 1–12 (2020)
27. Timimi, S., Milton, D., Bovell, V., Kapp, S., Russell, G.: Deconstructing diagnosis: four commentaries on a diagnostic tool to assess individuals for autism spectrum disorders. Autonomy (Birmingham, England) **1**(6), AR26 (2019)
28. Vaswani, A., et al.: Attention is all you need. In: Advances in Neural Information Processing Systems, vol. 30 (2017)
29. Wang, Y., Liu, J., Xiang, Y., Wang, J., Chen, Q., Chong, J.: MAGE: automatic diagnosis of autism spectrum disorders using multi-atlas graph convolutional networks and ensemble learning. Neurocomputing **469**, 346–353 (2022)
30. Wang, Y., Wang, J., Wu, F.X., Hayrat, R., Liu, J.: AIMAFE: autism spectrum disorder identification with multi-atlas deep feature representation and ensemble learning. J. Neurosci. Meth. **343**, 108840 (2020)
31. Yan, C., Zang, Y.: DPARSF: a MATLAB toolbox for "pipeline" data analysis of resting-state fMRI. Front. Syst. Neurosci. **4**, 1–17 (2010)
32. Zerveas, G., Jayaraman, S., Patel, D., Bhamidipaty, A., Eickhoff, C.: A transformer-based framework for multivariate time series representation learning. In: Proceedings of the 27th ACM SIGKDD Conference on Knowledge Discovery and Data Mining, August 2021. ACM (2021)
33. Zhao, Y., Ge, F., Zhang, S., Liu, T.: 3D deep convolutional neural network revealed the value of brain network overlap in differentiating autism spectrum disorder from healthy controls. In: Frangi, A.F., Schnabel, J.A., Davatzikos, C., Alberola-López, C., Fichtinger, G. (eds.) MICCAI 2018. LNCS, vol. 11072, pp. 172–180. Springer, Cham (2018). https://doi.org/10.1007/978-3-030-00931-1_20

Copy Number Variation Informs fMRI-Based Prediction of Autism Spectrum Disorder

Nicha C. Dvornek[1,2(✉)], Catherine Sullivan[3], James S. Duncan[1,2], and Abha R. Gupta[3]

[1] Department of Radiology and Biomedical Imaging, Yale School of Medicine, New Haven, CT 06510, USA
[2] Department of Biomedical Engineering, Yale University, New Haven, CT 06511, USA
[3] Department of Pediatrics, Yale School of Medicine, New Haven, CT 06510, USA
{nicha.dvornek,catherine.sullivan,james.duncan,abha.gupta}@yale.edu

Abstract. The multifactorial etiology of autism spectrum disorder (ASD) suggests that its study would benefit greatly from multimodal approaches that combine data from widely varying platforms, e.g., neuroimaging, genetics, and clinical characterization. Prior neuroimaging-genetic analyses often apply naive feature concatenation approaches in data-driven work or use the findings from one modality to guide posthoc analysis of another, missing the opportunity to analyze the paired multimodal data in a truly unified approach. In this paper, we develop a more integrative model for combining genetic, demographic, and neuroimaging data. Inspired by the influence of genotype on phenotype, we propose using an attention-based approach where the genetic data guides attention to neuroimaging features of importance for model prediction. The genetic data is derived from copy number variation parameters, while the neuroimaging data is from functional magnetic resonance imaging. We evaluate the proposed approach on ASD classification and severity prediction tasks, using a sex-balanced dataset of 228 ASD and typically developing subjects in a 10-fold cross-validation framework. We demonstrate that our attention-based model combining genetic information, demographic data, and functional magnetic resonance imaging results in superior prediction performance compared to other multimodal approaches.

Keywords: fMRI · Genetics · Multimodal analysis · Autism spectrum disorder

1 Introduction

Autism spectrum disorder (ASD) is characterized by impaired communication and social skills, and restricted, repetitive, and stereotyped behaviors that result in significant disability [2]. ASD refers to a spectrum of disorders due to its heterogeneity, with multiple etiologies, sub-types, and developmental trajectories [21], resulting in diverse clinical presentation of symptoms and severity. Two major factors contributing to the heterogeneity of ASD include *genetic variability*

A. Abdulkadir et al. (Eds.): MLCN 2023, LNCS 14312, pp. 133–142, 2023.
https://doi.org/10.1007/978-3-031-44858-4_13

and *sex* [21]. Research aimed at uncovering the pathophysiology of ASD and its heterogeneous presentations is critical to reduce disparities in early diagnosis and develop personalized targeted treatments.

A popular data-driven approach for discovering biomarkers for ASD is to first build a classification model that can distinguish ASD vs. typically developing (TD) individuals. Prior work generally focuses on unimodal data, e.g., functional magnetic resonance imaging (fMRI), structural MRI, genetics, or behavioral or developmental scores alone [13]. However, given the multifactorial etiology of ASD [19], a unified multimodal approach should improve model classification performance. Furthermore, convergence between different modality datasets on where in the brain ASD may arise and which brain regions correlate with different clinical measures would provide greater confidence in the results.

Prior multimodal methods combining genetic and neuroimaging data often naively concatenate the multimodal features and use them as inputs in a machine learning algorithm [4,11,27]. We aim to integrate the multimodal data in a more informative model design. Furthermore, such concatenation approaches may suffer from differing scales of the multimodal data, which will artificially give one modality greater importance. Another major direction is to use the findings from one modality to guide posthoc analysis of another [3,14,26], missing the opportunity to analyze the paired multimodal data in a truly unified approach. While phenotypic data has been combined with neuroimaging data in multiple ASD studies using thoughtful model designs [8,10,22], such integration between genetic and neuroimaging data has not been explored.

Here, we propose to improve characterization of the neurobiology of ASD by developing an integrated neuroimaging-genetic deep learning model to accurately classify ASD vs. TD individuals and predict ASD severity. As individual genetic differences will influence neuroimaging phenotypes, we propose using an attention-based approach where genetic variables inform what neuroimaging features should be attended to for model prediction. We assess the performance of the proposed approach on ASD classification and severity prediction tasks, using a sex-balanced dataset of 228 ASD and TD subjects in a 10-fold cross-validation framework. We demonstrate superior performance to other methods of combining multimodal data.

2   Methods

2.1   Dataset and Preprocessing

The dataset includes a sex-balanced cohort of 228 youth (age range: 8.0–17.9 years), 114 with ASD (59 female, 55 male) and 114 TD controls (58 female, 56 male), publicly available from NIH NDA collection 2021[1]. Data types utilized include clinical measures, genome-wide genotyping, and neuroimaging.

Clinical measures utilized include age at time of scan, sex, ASD diagnosis, and Social Responsiveness Scale-2 (SRS, range: 1–162) [7]. SRS is a measure of

---

[1] https://nda.nih.gov/edit_collection.html?id=2021

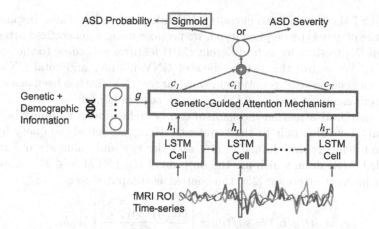

Fig. 1. The proposed integrated neuroimaging-genetic model. Genetic and demographic information will be used to focus attention to different fMRI features for the prediction task.

severity of social impairment in ASD, but was assessed on both ASD and TD subjects. Age and sex were used as demographic input variables to the models, while ASD status and SRS were used as prediction targets.

Genome-wide genotyping data was generated using the HumanOmni 2.5M BeadChip (Illumina). This data was processed and analyzed for rare (>50% of copy number variants (CNV) at <1% frequency in the Database of Genomic Variants) genic CNVs [14]. CNV parameters of number of CNVs (range: 0–10) and summed total length of all CNVs (range: 0–4050194 bp) were used as model inputs.

Structural and fMRI data were acquired under several different tasks. We utilized the Biopoint task-fMRI acquisition, in which subjects viewed coherent and scrambled biological point-light animations in alternating blocks (24 s per block; 154 volumes; TR = 2000 ms; TE = 30 ms; flip angle = 90°; FOV = 192 mm; image matrix = 64 mm^2; voxel size = 3 mm × 3 mm × 4 mm; 34 slices). Prior Biopoint studies have identified dysfunction in biological motion processing as reflecting key neural signatures of ASD and as a neuroendophenotype of genetic risk in unaffected siblings [5,17]. fMRI data was preprocessed in FSL [16], including MCFLIRT motion correction, interleaved slice timing correction, BET brain extraction, spatial smoothing (FWHM 5mm), high-pass temporal filtering, and registration to Montreal Neurological Institute space. We parcellated the brain into 116 regions of interest (ROIs) with the AAL atlas [24]. Standardized mean time-series from each ROI were used as model inputs.

2.2   Network Architecture

The structure of the proposed genetic-neuroimaging model is shown in Fig. 1. The fMRI time-series from predefined ROIs is first input to a long short-term memory (LSTM) layer [12] to encode fMRI information [9]. Then, inspired by the influence of genotype on phenotype, we propose using a generalized attention mechanism [25] to steer focus to different fMRI features according to the genetic information. We utilize the genetic data of CNV number and total CNV size derived from genome-wide genotyping, since larger CNV size has been associated with deleteriousness in previous ASD genetics studies [15,23].

The generalized attention mechanism can be defined as a mapping between a query and key-value pair to the "context". Here, we define the query by the genetic and demographic data $g \in R^G$, and the key and value are defined by the encoded fMRI data, which are the outputs of the LSTM $h_t \in R^L$. Applying scaled dot product attention [25], the context is computed as

$$c_t = att(g, h_t) = softmax \left[\frac{(W_q g)^T (W_k h_t)}{\sqrt{M}} \right] W_v h_t, \tag{1}$$

where $softmax(a_t)$ normalizes a_t such that $\sum_{t=1}^{T} a_t = 1$, W_q encodes the genetic and demographic information g into the query, W_k and W_v encode fMRI-based h_t into the key and value, respectively, and M denotes the dimension of the encoded space. Multiple attention mappings with different encodings W could be learned to diversify ways in which genetic information modulates the fMRI information. We then summarize the context across the T timesteps $\sum_{t=1}^{T} c_t$ and apply a fully connected layer to predict the output from the summary M context features. The ASD classification model ends with a sigmoid activation function to produce the probability of the ASD class, while the ASD severity regression model ends with the output of the fully connected layer.

2.3 Experimental Settings

To assess the effectiveness of our multimodal attention-based approach (Att), we compared to 4 other methods:
1) $Base$: A basic LSTM model using only the fMRI time-series data [9].
2) $Concat$: Concatenation of the genetic and demographic data with the fMRI time-series data. The genetic and demographic values are repeated across time and concatenated as additional features to the fMRI data. The data are then input to the basic LSTM model.
3) $Fusion$: Fusion of the genetic and demographic information with the prediction from the LSTM processed fMRI data [8]. The genetic and demographic information is combined with the LSTM block using a fully connected layer, followed by a layer with a single node for the task prediction.
4) $LSTMinit$: The baseline LSTM model with initialization of hidden states conditioned on the genetic and demographic variables [10]. The LSTM initial state vectors are learned using a single fully connected layer with size L.

We also perform an ablation study to assess the utility of including both genetic and demographic information to guide the attention to fMRI features in the proposed full model. We thus trained reduced models that included only demographic (*Att-demo*) or genetic (*Att-gene*) information alone to evaluate their value.

All models were implemented in Python using Keras [6] and Tensorflow [1] libraries. The feature dimension for the LSTM output was $L = 16$, and the feature dimension for the attention embedding was set to $M = 8$. fMRI time-series for each ROI were standardized to have mean 0 and standard deviation 1, and genetic and demographic inputs were normalized to the range $[-1, 1]$. SRS raw scores were normalized to the range $[0, 1]$. We applied randomly shifted windowed samples of the fMRI time-series data as data augmentation [9], sampling 10 random windows of size $T = 48$ (representing the time for 2 blocks each of scrambled and biological motion) per subject every epoch. We used the Adam optimizer [18] ($lr = 0.001$) to train the models for up to 50 epochs. Classification models were trained to optimize binary cross-entropy loss and severity regression models were trained to optimize mean squared error loss.

To evaluate model performance, we performed 10-fold cross-validation of subjects, using stratified sampling of ASD status, with 80% subjects for training, 10% for validation, and 10% for testing in each partition. Validation loss was used to determine the stopping epoch for model selection. For classification models, we computed classification accuracy, sensitivity, specificity, and area under the receiver operating characteristic curve (AUC) for each fold. We also computed the overall AUC based on aggregating the test predictions from all folds. For regression models, we computed the mean squared error (MSE), maximum of the mean squared error, and Pearson correlation between true and predicted test outputs for each fold. Because Pearson correlation can vary largely when there are small perturbations in a small test dataset, we also computed the overall Pearson correlation based on aggregating the test predictions from all folds. Significant differences between models were assessed using paired two-tailed t-tests to compare matched cross-validation folds ($\alpha = 0.05$).

3   Results

3.1   ASD Classification

Classification Performance. ASD classification model results are summarized in Table 1, and receiver operating characteristic curves are plotted in Fig. 2. Traditional concatenation of multimodal inputs performed worse than the fMRI-only base model. Fusing multimodal features in later layers produced the highest sensitivity, but was not significantly different from the base or our attention-based model results. Initializing the LSTM with genetic and demographic information performed similarly to the fMRI-only base model. Our attention-based approach using the genetic information to inform the attending of important fMRI features resulted in the highest classification accuracy, specificity, and AUC. Furthermore, our attention-based model is the only multimodal approach to perform significantly better than the fMRI-only base model based on the accuracy metric. Moreover, our approach significantly outperformed the other multimodal models as measured by accuracy, specificity, and AUC. As seen in Fig. 2, the receiver operating characteristic curve for our attention model lies above the other models for much of the plot.

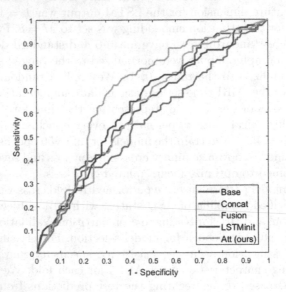

Fig. 2. Receiver operating characteristic curves from ASD classification models.

When training the attention-based model with only demographic or genetic data for guidance, the models still produced generally better results than the non-attention models, but with a small performance drop compared to the full model utilizing both demographic and genetic information. Using genetic information alone resulted in significantly worse specificity compared to the full model. Again, only the full model utilizing both genetic and demographic data performed significantly better than the unimodal fMRI base model (as measured by accuracy), suggesting the importance of including all modalities in computing the attention scores.

Model Analysis. We investigated our attention-based model's representation of multimodal information via the summary context vector. First, we performed a tSNE visualization [20] of the context for one trained model (Fig. 3), where TD subjects are plotted in red and ASD subjects are plotted in blue. We see that while overlap still exists, there is clustering of the TD and ASD samples.

We then investigate the impact of each genetic and demographic variable on each context feature. We compute the Pearson correlation between a genetic or demographic variable $g(i)$ and the ASD diagnosis and each feature of the context representation $c(j)$ across all samples in a representative fold. We visualize the input and output features with significant correlations ($p < 0.05$) for each context feature by white boxes in Fig. 4. As expected, every context feature is

Table 1. ASD classification performance (mean ± standard deviation). Best results marked in bold.

Model	Accuracy	Sensitivity	Specificity	AUC	Overall AUC
Base	0.59 ± 0.13^a	0.67 ± 0.19	0.52 ± 0.22	0.63 ± 0.14	0.62
Concat	0.54 ± 0.07^a	0.64 ± 0.14	0.44 ± 0.15^a	0.56 ± 0.12^{ab}	0.56
Fusion	0.60 ± 0.12^a	$\mathbf{0.73 \pm 0.11}$	0.46 ± 0.20^a	0.63 ± 0.13	0.62
LSTMinit	0.62 ± 0.11	0.69 ± 0.12	0.54 ± 0.12^a	0.64 ± 0.14	0.63
Att-demo	0.68 ± 0.06	0.67 ± 0.13	0.68 ± 0.11	$\mathbf{0.70 \pm 0.08}$	0.69
Att-gene	0.67 ± 0.06	0.68 ± 0.11	0.65 ± 0.13^a	0.69 ± 0.08	0.69
Att (ours)	$\mathbf{0.69 \pm 0.06}^b$	0.68 ± 0.13	$\mathbf{0.69 \pm 0.11}$	$\mathbf{0.70 \pm 0.08}$	**0.70**

[a] Significant difference compared to our proposed approach Att ($p < 0.05$, paired two-tailed t-test)

[b] Significant difference compared to fMRI-only Base model ($p < 0.05$, paired two-tailed t-test)

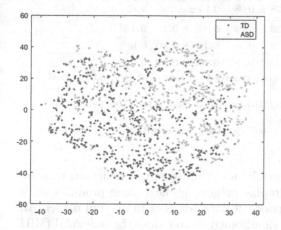

Fig. 3. tSNE visualization of the summary context vector for each sample.

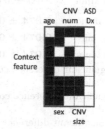

Fig. 4. White boxes denote significant correlations ($p < 0.05$) between model context features (rows) and different genetic and demographic features (columns).

correlated with the ASD diagnosis. Furthermore, we see that different context features are guided by different input features. For example, the context feature corresponding to row 1 has only 1 feature, i.e., age, associated with it, while the second context feature is associated with both genetic variables. Our attention-based approach allows for different modalities of inputs to be integrated in a dynamic manner to produce more informative, individualized representations of the input data, which lends to the improved classification performance.

3.2   ASD Severity Regression

ASD severity regression results are summarized in Table 2. The fMRI-only base model was able to achieve a significant overall correlation. Concatenating genetic and demographic features with fMRI nominally improved the prediction performance. Fusing the genetic and demographic features with fMRI features significantly reduced the MSE compared to the fMRI-only base model and produced the lowest maximum errors; however, the correlation between predicted and true scores was greatly reduced and no longer significant. The model initializing the LSTM with genetic and demographic variables performed slightly worse than the base fMRI model. Our attention-based approach resulted in the lowest MSE and highest correlation.

Table 2. ASD severity prediction performance (normalized SRS mean ± standard deviation). Best results marked in bold.

Model	MSE	Maximum	Correlation	Overall Correlation
Base	0.084 ± 0.014^a	0.30 ± 0.09	0.16 ± 0.19	0.15^c
Concat	0.080 ± 0.016	0.29 ± 0.06	0.17 ± 0.21	0.16^c
Fusion	0.075 ± 0.016^b	$\mathbf{0.26 \pm 0.08}$	0.08 ± 0.12	0.08
LSTMinit	0.087 ± 0.011	0.28 ± 0.08	0.13 ± 0.28	0.12
Att-demo	$\mathbf{0.074 \pm 0.019}^b$	0.29 ± 0.09	$\mathbf{0.19 \pm 0.11}$	$\mathbf{0.18}^d$
Att-gene	$\mathbf{0.074 \pm 0.019}^b$	0.29 ± 0.08	$\mathbf{0.19 \pm 0.10}$	0.16^c
Att (ours)	$\mathbf{0.074 \pm 0.018}^b$	0.29 ± 0.09	$\mathbf{0.19 \pm 0.12}$	$\mathbf{0.18}^d$

[a] Significant difference compared to our proposed approach Att ($p < 0.05$, paired two-tailed t-test)
[b] Significant difference compared to fMRI-only Base model ($p < 0.05$, paired two-tailed t-test)
[c]/[d] Significant difference from 0 ($p < 0.05/p < 0.01$, two-tailed t-test)

As seen in the ablation study results for ASD severity prediction, training the attention module with demographic or genetic data alone produces similar results to training with both modes of data together in our full model. All attention-based models performed significantly better than the unimodal fMRI model as measured by MSE. Using genetic data alone produced a slightly lower overall correlation. Although the performance levels are similar, including both genetic and demographic information may result in better understand of the interplay between these different types of individual features in the context of understanding ASD.

4   Conclusions

In this work, we proposed an attention-based approach to integrating genetic and demographic information with fMRI data. The genetic and demographic data are used to guide attention to important fMRI encoded features. In a 10-fold cross-validation framework with a dataset of 228 ASD and TD subjects, we demonstrated improved performance in an ASD classification and ASD severity prediction task compared to other standard approaches of combining multimodal data. While acquiring such a multimodal imaging and genetic dataset from a unique cohort requires extensive time and resources, we acknowledge that the generally smaller dataset used in this study is a limitation of the presented validation. Future work will verify the model performance on larger datasets, investigate additional sequence modeling methods such as transformers [25], explore other genetic variables to use as input (e.g., number of genes contained in CNVs), and analyze the neuroimaging biomarkers that are used by the attention-based model for ASD classification and severity prediction.

Acknowledgements. This work was supported in part by the National Institute of Health grant R01NS035193.

References

1. Abadi, M., et al.: TensorFlow: large-scale machine learning on heterogeneous systems (2015). http://www.tensorflow.org/, software available from tensorflow.org
2. American Psychiatric Association, Washington, DC: Diagnostic and Statistical Manual of Mental Disorders, 5th edn. (2013)
3. Antonucci, L.A., et al.: Thalamic connectivity measured with fMRI is associated with a polygenic index predicting thalamo-prefrontal gene co-expression. Brain Struct. Funct. **224**, 1331–1344 (2019)
4. Bi, X., Hu, X., Xie, Y., Wu, H.: A novel CERNNE approach for predicting Parkinson's disease-associated genes and brain regions based on multimodal imaging genetics data. Med. Image Anal. **67**, 101830 (2021)
5. Björnsdotter, M., Wang, N., Pelphrey, K., Kaiser, M.D.: Evaluation of quantified social perception circuit activity as a neurobiological marker of autism spectrum disorder. JAMA Psychiat. **73**(6), 614–621 (2016)
6. Chollet, F., et al.: Keras (2015). Software available on Github
7. Constantino, J.N., Gruber, C.P.: Social Responsiveness Scale: SRS-2. Western Psychological Services Torrance, CA (2012)
8. Dvornek, N.C., Ventola, P., Duncan, J.S.: Combining phenotypic and resting-state fMRI data for autism classification with recurrent neural networks. In: 2018 IEEE 15th International Symposium on Biomedical Imaging, ISBI 2018, pp. 725–728. IEEE (2018)
9. Dvornek, N.C., Ventola, P., Pelphrey, K.A., Duncan, J.S.: Identifying autism from resting-state fMRI using long short-term memory networks. In: Wang, Q., Shi, Y., Suk, H.-I., Suzuki, K. (eds.) MLMI 2017. LNCS, vol. 10541, pp. 362–370. Springer, Cham (2017). https://doi.org/10.1007/978-3-319-67389-9_42
10. Dvornek, N.C., Yang, D., Ventola, P., Duncan, J.S.: Learning generalizable recurrent neural networks from small task-fMRI datasets. In: Frangi, A.F., Schnabel, J.A., Davatzikos, C., Alberola-López, C., Fichtinger, G. (eds.) MICCAI 2018. LNCS, vol. 11072, pp. 329–337. Springer, Cham (2018). https://doi.org/10.1007/978-3-030-00931-1_38
11. Heinrich, A., et al.: Prediction of alcohol drinking in adolescents: personality-traits, behavior, brain responses, and genetic variations in the context of reward sensitivity. Biol. Psychol. **118**, 79–87 (2016)
12. Hochreiter, S., Schmidhuber, J.: Long short-term memory. Neural Comput. **9**(8), 1735–1780 (1997)
13. Hyde, K.K., et al.: Applications of supervised machine learning in autism spectrum disorder research: a review. Rev. J. Autism Dev. Disord. **6**, 128–146 (2019)

14. Jack, A., et al.: A neurogenetic analysis of female autism. Brain **144**(6), 1911–1926 (2021)
15. Jacquemont, S., et al.: A higher mutational burden in females supports a "female protective model" in neurodevelopmental disorders. Am. J. Hum. Genet. **94**(3), 415–425 (2014)
16. Jenkinson, M., Beckmann, C.F., Behrens, T.E., Woolrich, M.W., Smith, S.M.: FSL. Neuroimage **62**(2), 782–790 (2012)
17. Kaiser, M.D., et al.: Neural signatures of autism. Proc. Natl. Acad. Sci. **107**(49), 21223–21228 (2010)
18. Kingma, D.P., Ba, J.: Adam: a method for stochastic optimization (2017)
19. Lord, C., et al.: Autism spectrum disorder. Nat. Rev. Dis. Primers. **6**(1), 1–23 (2020)
20. Van der Maaten, L., Hinton, G.: Visualizing data using t-SNE. J. Mach. Learn. Res. **9**(11), 2579–2605 (2008)
21. Masi, A., DeMayo, M.M., Glozier, N., Guastella, A.J.: An overview of autism spectrum disorder, heterogeneity and treatment options. Neurosci. Bull. **33**, 183–193 (2017)
22. Parisot, S., et al.: Spectral graph convolutions for population-based disease prediction. In: Descoteaux, M., Maier-Hein, L., Franz, A., Jannin, P., Collins, D.L., Duchesne, S. (eds.) MICCAI 2017. LNCS, vol. 10435, pp. 177–185. Springer, Cham (2017). https://doi.org/10.1007/978-3-319-66179-7_21
23. Pinto, D., et al.: Convergence of genes and cellular pathways dysregulated in autism spectrum disorders. Am. J. Hum. Genet. **94**(5), 677–694 (2014)
24. Tzourio-Mazoyer, N., et al.: Automated anatomical labeling of activations in SPM using a macroscopic anatomical parcellation of the MNI MRI single-subject brain. Neuroimage **15**(1), 273–289 (2002)
25. Vaswani, A., et al.: Attention is all you need. In: Neural Information Processing Systems (2017)
26. Vértes, P.E., et al.: Gene transcription profiles associated with inter-modular hubs and connection distance in human functional magnetic resonance imaging networks. Philos. Trans. R. Soc. B Biol. Sci. **371**(1705), 20150362 (2016)
27. Yang, H., Liu, J., Sui, J., Pearlson, G., Calhoun, V.D.: A hybrid machine learning method for fusing fMRI and genetic data: combining both improves classification of schizophrenia. Front. Hum. Neurosci. **4**, 192 (2010)

Deep Attention Assisted Multi-resolution Networks for the Segmentation of White Matter Hyperintensities in Postmortem MRI Scans

Anoop Benet Nirmala[1]([⊠]), Tanweer Rashid[1], Elyas Fadaee[1],
Nicolas Honnorat[1], Karl Li[1], Sokratis Charisis[1], Di Wang[1],
Aishwarya Vemula[1], Jinqi Li[1], Peter Fox[1], Timothy E. Richardson[2],
Jamie M. Walker[1], Kevin Bieniek[1], Sudha Seshadri[1], and Mohamad Habes[1]

[1] Glenn Biggs Institute for Alzheimer's and Neurodegenerative Diseases,
University of Texas Health Science Center San Antonio,
San Antonio, TX, USA
benetnirmala@uthscsa.edu
[2] Icahn School of Medicine at Mount Sinai, New York, USA

Abstract. In the presence of cardiovascular disease and neurodegenerative disorders, the white matter of the brains of clinical study participants often present bright spots in T2-weighted Magnetic Resonance Imaging scans. The pathways contributing to the emergence of these white matter hyperintensities are still debated. By offering the possibility to directly compare MRI patterns with cellular and tissue alterations, research studies coupling postmortem imaging with histological studies are the most likely to provide a satisfactory answer to these open questions. Unfortunately, manually segmenting white matter hyperintensities in postmortem MRI scans before histology is time-consuming and labor-intensive. In this work, we propose to tackle this issue with new, fully automatic segmentation tools relying on the most recent Deep Learning architectures. More specifically, we compare the ability to predict white matter hyperintensities from a registered pair of T1 and T2-weighted postmortem MRI scans of five Unet architectures: the original Unet, DoubleUNet, Attention UNet, Multiresolution UNet, and a new architecture specifically designed for the task. A detailed comparison between these five Unets and an ablation study, carried out on the sagittal slices of 13 pairs of high-resolution T1 and T2 weighted MRI scans manually annotated by neuroradiologists, demonstrate the superiority of our new approach and provide an estimation of the performance gains offered by the modules introduced in the new architecture.

Keywords: Convolutional Neural Network · White Matter Hyperintensities · Postmortem Brain MRI

1 Introduction

White matter hyperintensities (WMH) are aging traits that are frequently considered a subtype of cerebral small vessel disease (SVD) [10,11]. Their prevalence increases with age and ends up affecting the majority of the population [32].

A. Abdulkadir et al. (Eds.): MLCN 2023, LNCS 14312, pp. 143–152, 2023.
https://doi.org/10.1007/978-3-031-44858-4_14

White matter hyperintensities can be clinically identified by Magnetic Resonance Imaging (MRI) because they appear as bright spots in T2-weighted MRI images [25, 30]. White matter hyperintensities are observed in normal aging and in the presence of vascular dementia, but they have also been identified as a key feature of Alzheimer's Disease [17]. White matter hyperintensities have been the subject of numerous in-vivo clinical studies [10, 11, 17]. Unfortunately, these studies could not fully elucidate the cellular and brain tissue alterations at the origin of the MRI intensity shifts observed [16]. Postmortem brain MRI studies emerged as a promising approach to tackle this issue [16] and, in particular, because working with postmortem brain samples fixed in formalin offers the possibility to conduct neuropathological analyses on brain tissues extracted exactly within the WMH regions visible in the MRI scans [9, 20]. Unfortunately, the manual WMH annotations that the neurologists would need to conduct to locate interesting brain regions for histology are time-consuming to obtain [4, 28].

In this study, we propose to tackle this issue by building upon the automatic and semi-automatic MRI segmentation tools that were proposed for detecting WMH inside in-vivo MRI scans [1, 6, 7, 18, 26, 27]. The first approaches adopted to delineate WMH rely on binary segmentation, clustering, and semi-automatic approaches. For instance, Samaille et al. used the watershed segmentation method to directly segment WMH [26]. Anbeek et al. proposed a K nearest-neighbor classification method and used T1-weighted and T2-weighted MRI to estimate the likelihood for a brain location to belong to a region of white matter hyperintensities [1]. Several clustering algorithms were used for segmenting WMH, such as the fuzzy c-means clustering algorithm [7] and Gaussian mixture models [27]. The statistical Parametric Mapping (SPM) Lesion Segmentation Toolbox was used by Maldjian et al. for segmenting WMH [18] and, more recently, Fiford et al. proposed a semi-automatic approach based on Bayesian Model Selection [6]. The recent emergence of Deep Learning models achieving unprecedented performances for image segmentation tasks [13, 15, 21, 24] led to a new generation of approaches for brain lesions segmentation [23] and WMH delineation [8]. The supervised deep learning architecture Unet [24], in particular, produced remarkable results for the task [31]. But unfortunately, none of these recent models was used to delineate WMH in postmortem scans. Further experiments are still required to check how these new tools would handle the specific patterns and intensity ranges of postmortem scans.

The present work is intended to address this gap. More specifically, we compare the performance of four variants of Unet, the most established deep network architecture for image segmentation: the original Unet [24], the DoubleUNet [15], the Attention UNet [21], and the Multi-resolution UNet [13] with a new Unet architecture specifically designed to extract more accurate WMH segmentations from pairs of registered T1 and T2-weighted postmortem MRI scans. The new architecture proposed in this work implements attention-gated skip connections [21] and a multi-scale module based on atrous spatial pyramid pooling [5] to produce more accurate segmentations, and will therefore be denoted the deep attention assisted multi-resolution (deep AIM) architecture. We report detailed

raw scan   pre-processed scan   annotations

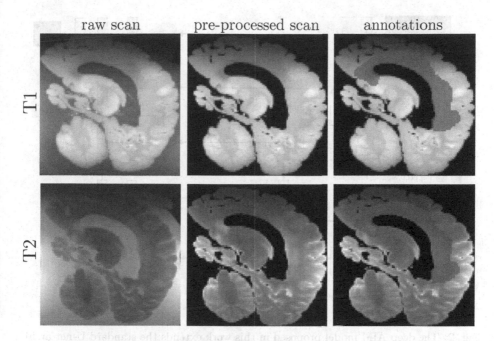

Fig. 1. Postmortem T1-weighted and T2-weighted MRI scans were segmented to remove the formalin signal and fix magnetic field inhomogeneities. This pre-processing greatly helped the manual annotation of white matter lesions.

cross-validated comparison and ablation studies carried out on the sagittal slices of 13 pairs of postmortem T1 and T2 weighted MRI scans manually annotated by neuroradiologists. Our results suggest that deep AIM produces more accurate WMH maps than standard Unets, and provide an estimation of the performance improvements granted by the attention gates and the multi-scale module.

2   Methods

2.1   Data and Pre-processing

Thirteen brains donated to the brain bank of the University of Texas Health Science Center at San Antonio and fixed in formalin were considered in this study [2]. For each brain, a T1-weighted and a T2-weighted MRI scans of the whole left hemisphere were conducted. The T1-weighted scans were acquired at a resolution of $0.5 \times 0.5 \times 0.5$ mm, repetition time of 2.2 s, echo time of 3.27 ms, flip angle 3°. The T2 scans were acquired at a resolution of $0.625 \times 0.625 \times 1.5$ mm, repetition time of 6.96 s, echo time of 68 ms, flip angle 120°. All the scans were manually re-oriented in the same direction. They were denoised by conducting a N4 bias field correction [29] and applying the non-local mean filtering implemented in NAONLM3D [19]. The denoised T2 scans were then interpolated to an isotropic 0.5 mm resolution using the ResampleImage function provided with ANTS (version 2.1.0-gGIT-N) set for a windowedSinc interpolation [3]. The resampled T2

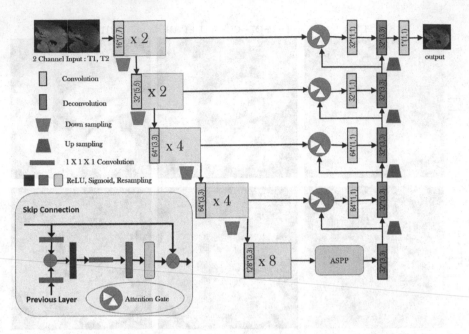

Fig. 2. The deep AIM model proposed in this work extends the standard U-net architecture by applying an Atrous Spatial Pyramid Pooling (ASPP) [5] module to extract better high-level features and filtering the skip connections with attention gates to produce more accurate segmentations.

scans were registered to the T1 scans by using the ANTS rigid registration set to optimize mutual information [3]. A brain mask was generated by FSL BET, manually fixed, and then applied to mask out the formalin signal [14]. Lastly, the image intensities of the masked denoised T1 and the registered T2 scans were linearly scaled, independently for each image, to set the smallest intensity value to 0 and the largest intensity value to 1. Two board-certified neuroradiologists manually segmented the registered T2 scans to delineate the brain regions with WMH. The final T1, registered T2, and manual mask were all of size $174 \times 338 \times 478$. The 174 sagittal slices of these scans were all individually resized to 128×128 using the imresize command of the opencv python library (version 4.5.5.64), and we trained our deep networks to predict the resized sagittal slices of the manual WMH masks from the corresponding pairs of resized T1 and T2 MRI slices. Figure 1 illustrates this pre-processing by presenting the original, pre-processed, and annotated scans corresponding to the first brain sample considered in this study.

2.2 UNet

Five Unet architectures were investigated in this work: a standard Unet [24], a DoubleUNet (D_Unet) [15], an Attention UNet (A_Unet) [21], a Multi-resolution

UNet (M_Unet) [13], and the new architecture presented in the following section that was designed to leverage the advantages of these four architectures.

UNet is a Convolutional Neural Network (CNN) architecture that has established a new standard for image segmentation [24]. The network derives its name from its U-shaped architecture made of an encoder and a decoder connected by skip connections. The encoder captures high-level features through a series of convolutional and pooling layers, while the decoder gradually upsamples the features to generate a segmentation map that matches the size of the input image. The skip connections feed each layer of the decoder with the output of the encoder layers corresponding to the same image resolution to combine low-level and high-level features and produce significantly more accurate segmentations than a simple autoencoder architecture [24].

DoubleUNet (D_Unet) is an extension of the UNet architecture that stacks two UNet models to form a more powerful network [15]. DoubleUNet leverages the hierarchical representations learned by each UNet model to enhance the segmentation performance with the help of the Atrous Spatial Pyramid Pooling (ASPP) layer [5,15].

Attention UNet (A_Unet) is a variant of the UNet architecture that incorporates attention mechanisms to enhance the performance of image segmentation tasks [21]. The attention mechanism allows the network to selectively focus on informative regions of the input image while suppressing irrelevant or noisy regions. The attention mechanism in an Attention UNet typically consists of two main components: the query, key, and value (QKV) representations and the attention map. The QKV representations are generated from the feature maps of the encoder, and they are used to compute the attention map. The attention map represents the spatial importance of each feature map element. It can be seen as a weight map that determines how much attention should be given to each element when decoding. The attention map is then combined with the feature maps in the decoder part to guide the upsampling process. This attention mechanism lets the network focus on relevant features and suppress irrelevant or noisy features, which produces more accurate segmentation results.

Multi-resolution UNet (M_Unet) is a variant of UNet that fuses feature maps from multiple resolutions to improve the image segmentation [13]. In a standard UNet, the encoder part of the network captures high-level features through a series of convolutional and pooling layers, while the decoder part gradually upsamples the features to generate the segmentation map. In a Multi-resolution UNet, additional pathways are introduced to fuse feature maps from different resolutions in the decoder part. This fusion process improves the combination of local and global information during the upsampling process, which produces more robust and accurate results.

2.3 Proposed Deep AIM Architecture

The architecture of the deep AIM model proposed in this work is depicted in Fig. 2. The overall structure of the network, made of five sets of encoding convolutional layers of decreasing kernel sizes and five sets of decoding layers linked by

Table 1. Average and standard deviation of metrics measured during the cross-validation, comparing the ground truth manual WMH segmentations and the maps generated by the five variants of Unet investigated in this work, with or without data augmentation (DA). The accuracy is a balanced accuracy.

Models	Dice	Precision	Recall	Accuracy	Specificity
Unet [24]	58.14(19.43)	**81.05(24.96)**	49.70(15.10)	74.81(14.92)	**99.93(0.09)**
Unet+DA [24]	65.67(17.35)	60.13(20.69)	80.56(11.26)	90.12(11.29)	99.69(00.22)
D_Unet [15]	46.82(17.72)	34.41(15.70)	87.75(09.62)	93.27(09.98)	98.79(00.96)
D_Unet+DA [15]	61.85(18.22)	63.56(22.12)	68.14(14.61)	83.95(00.22)	99.76(00.21)
A_Unet [21]	46.31(16.93)	33.87(15.12)	89.12(10.39)	93.93(10.06)	98.75(00.98)
A_Unet+DA [21]	65.27(23.43)	72.68(26.53)	65.71(19.85)	82.79(18.31)	99.87(00.14)
M_Unet [13]	62.09(16.87)	50.50(17.41)	90.01(08.78)	94.70(08.57)	99.39(00.55)
M_Unet+DA [13]	74.41(16.81)	83.22(21.53)	71.69(12.27)	85.80(11.12)	99.91(00.12)
deep AIM	66.43(03.45)	77.34(24.14)	65.31(15.10)	75.48(00.17)	99.39(00.57)
deep AIM+DA	**79.02(1.56)**	68.77(05.69)	**92.87(8.30)**	**96.22(0.19)**	99.58(00.18)

Table 2. Depp AIM ablation study. The accuracy is a balanced accuracy.

Models	Dice	Precision	Recall	Accuracy	Specificity
deep AIM without ASPP and Attention	64.16	73.72	59.58	78.89	99.49
deep AIM without ASPP	70.43	75.72	65.58	81.89	99.49
deep AIM without Attention	71.3	**76.83**	66.58	89.53	99.01
deep AIM	**79.02**	68.77	**92.87**	**96.22**	**99.58**

skip connections, is a variant of Unet structure [22,24], where we have placed an Atrous Spatial Pyramid Pooling (ASPP) layer on the deepest skip connection and filtered the other skip connections with attention modules [5,21]. During the decoding stage, these attention modules draw attention to pertinent features coming from the encoder side, while the ASPP module extracts high-level features at multiple scales to provide more information to the decoding layers. All the convolutional layers were implemented with batch normalization and ReLU activation. The ASPP module was implemented for the two dilation rates of 6 and 12 [5]. The attention modules use self-attention to jointly extract spatial and channel information [21]. The attention coefficients are normalized by performing a channel-wise $1 \times 1 \times 1$ convolution on both inputs, followed by a sigmoid activation function. The attention coefficients emphasize the pertinent features that are sent by the encoder to the decoder.

3   Experiments

The Unet models were all implemented using the python library Keras with Tensorflow as the backend, initialized using the *he normal* approach [12], and trained for 250 epochs using the Adam optimizer with learning rate 0.0003 and default parameters, for the Dice Loss and a batch size of 8. The computational times were measured on an NVIDIA DGX station with a 64-bit Ubuntu operating system, a 32 GB dedicated NVIDIA Tesla V100 GPU.

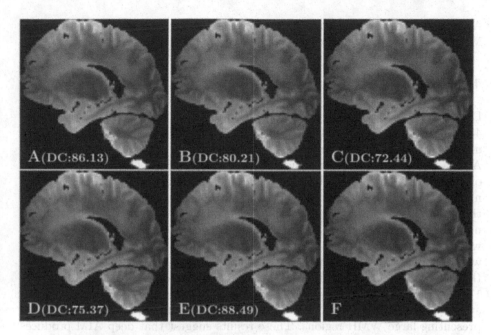

Fig. 3. WMH maps generated for the first pair of scans by Multi-resolution UNet (A), Unet (B), Double Unet (C), Attention Unet (D), and deep AIM method (E), and their Dice overlap (DC) with the ground truth manual segmentation (F).

The performances of the five Unet architectures were compared by conducting a Monte Carlo cross-validation. More specifically, the 13 pairs of MRI scans were randomly shuffled, the sagittal slices of the first 7 pairs and their manual segmentations were used to train the models, the sagittal slices of 3 pairs of scans were passed to the Keras optimizer for validation, and the validated models were applied to the sagittal slices of the remaining 3 pairs of scans to generate WMH maps. These WMH maps were compared with the associated ground truth manual segmentations by computing five metrics: Dice overlap (DC), precision, recall, balanced accuracy, and specificity. This process was repeated ten times, and the metrics were averaged over these ten repeats.

All the experiments were replicated with real-time data augmentation. Real-time data augmentation consists in randomly altering the training data during the training of a deep network. We used the data augmentation implemented in the ImageDataGenerator of the keras-preprocessing python library (version 1.1.2), which was set to a rotation angle of 90, width and height shift range of 0.3, shear range of 0.5, zoom range of 0.3, horizontal and vertical flips, and a fill mode set to *reflect*.

Lastly, an ablation study was conducted by comparing our proposed architecture, deep AIM, with an architecture where both the ASPP and the attention modules had been removed, a deep AIM architecture without ASPP, and a deep

AIM architecture without attention modules. During this ablation study, the Monte-Carlo cross-validation was conducted for five repeats.

4   Results

Deep AIM has 33,838,907 parameters to optimize and required on average 24 h 45 min 20 s for training and 5.82 s for testing. Table 1 reports the segmentation metrics measured for all the models, with and without data augmentation. Deep AIM achieved the best cross-validated Dice score and the best recall, while the original Unet achieved the best precision and specificity. Deep AIM is the only model to achieve a balanced accuracy larger than 95% (96.22%). The results of the ablation study presented in Table 2 indicate that both ASPP and attention modules contributed to this success. The balanced accuracy dropped by 6.7% when the attention modules were removed, and by 14.3% when the ASPP module was removed. Overall, removing all the modules proposed in this work reduced the balanced accuracy of the maps generated by the deep AIM architecture by 17.3%. Figure 3 illustrates these results by presenting the WMH maps generated by the different Unet architectures for the same postmortem brain slice presenting large WMH regions. These results suggest that deep AIM produces more accurate WMH maps than standard Unets, better capturing the regions annotated by neurologists. With a balanced accuracy above ninety-five percent, our proposed architecture generates maps that are already reliable enough to guide histology. We hope to reach an even better accuracy in the future by combining the WMH maps generated by deep AIM networks trained to segment axial, sagittal, and coronal slices and by extending the deep AIM field of view to process MRI scans in their original resolution.

5   Conclusion

In this study, we investigate the ability of Unet architectures to detect white matter hyperintensities in postmortem brain samples by jointly segmenting pairs registered of T1 and T2-weighted MRI scans, tuned to study postmortem brain tissues. In addition to four major architectures, we measure the performance of a new Unet architecture where we have introduced atrous spatial pyramid pooling and attention-gated modules to produce better segmentations, and we carry out an ablation study to measure the performance increases granted by these modules. Our results, derived via Monte Carlo cross-validation for a set of thirteen pairs of postmortem scans manually annotated by neuroradiologists, suggest that our new architecture produces a better map than standard Unets, and that all the modules introduced contributed to that success. The balanced accuracy of 96.22% obtained with data augmentation by our new architecture opens new research perspectives and, in particular, the possibility to segment white matter intensities in large sets of postmortem MRI scans, to study the nature and the origin of these brain lesions.

References

1. Anbeek, P., Vincken, K.L., Van Osch, M.J.P., Bisschops, R.H.C., Van Det Grond, J.: Probabilistic segmentation of white matter lesions in MR imaging. NeuroImage **21**(3), 1037–1044 (2004)
2. Anonymous: in revision
3. Avants, B., Tustison, N., Wu, J., Cook, P., Gee, J.: An open source multivariate framework for n-tissue segmentation with evaluation on public data. Neuroinformatics **9**(4), 381–400 (2011)
4. Benson, R.R., et al.: Older people with impaired mobility have specific loci of periventricular abnormality on MRI. Neurology **58**(1), 48–55 (2002)
5. Chen, L.-C., Zhu, Y., Papandreou, G., Schroff, F., Adam, H.: Encoder-decoder with atrous separable convolution for semantic image segmentation. In: Ferrari, V., Hebert, M., Sminchisescu, C., Weiss, Y. (eds.) ECCV 2018. LNCS, vol. 11211, pp. 833–851. Springer, Cham (2018). https://doi.org/10.1007/978-3-030-01234-2_49
6. Fiford, C.M., et al.: Automated white matter hyperintensity segmentation using bayesian model selection: assessment and correlations with cognitive change. Neuroinformatics **18**, 429–449 (2020)
7. Gibson, E., Gao, F., Black, S.E., Lobaugh, N.J.: Automatic segmentation of white matter hyperintensities in the elderly using flair images at 3t. J. Magn. Reson. Imaging **31**(6), 1311–1322 (2010)
8. Giese, A.K., et al.: White matter hyperintensity burden in acute stroke patients differs by ischemic stroke subtype. Neurology **95**(1), e79–e88 (2020)
9. Grinberg, L., et al.: Improved detection of incipient vascular changes by a biotechnological platform combining post mortem MRI in situ with neuropathology. J. Neurol. Sci. **283**(1–2), 2–8 (2009)
10. Habes, M., et al.: White matter hyperintensities and imaging patterns of brain ageing in the general population. Brain **139**(4), 1164–1179 (2016)
11. Habes, M., et al.: White matter lesions: spatial heterogeneity, links to risk factors, cognition, genetics, and atrophy. Neurology **91**(10), e964–e975 (2018)
12. He, K., Zhang, X., Ren, S., Sun, J.: Delving deep into rectifiers: surpassing human-level performance on imagenet classification. In: Proceedings of the IEEE International Conference on Computer Vision, pp. 1026–1034 (2015)
13. Ibtehaz, N., Rahman, M.S.: Multiresunet: rethinking the u-net architecture for multimodal biomedical image segmentation. Neural Netw. **121**, 74–87 (2020)
14. Jenkinson, M., Beckmann, C., Behrens, T., Woolrich, M., Smith, S.: FSL. NeuroImage **62**(2), 782–790 (2012)
15. Jha, D., Riegler, M.A., Johansen, D., Halvorsen, P., Johansen, H.D.: Doubleu-net: a deep convolutional neural network for medical image segmentation. In: 2020 IEEE 33rd International Symposium on Computer-Based Medical Systems (CBMS), pp. 558–564. IEEE (2020)
16. Jonkman, L.E., Kenkhuis, B., Geurts, J.J., van de Berg, W.D.: Post-mortem MRI and histopathology in neurologic disease: a translational approach. Neurosci. Bull. **35**, 229–243 (2019)
17. Lee, S., et al.: White matter hyperintensities are a core feature of Alzheimer's disease: evidence from the dominantly inherited Alzheimer network. Ann. Neurol. **79**(6), 929–939 (2016)
18. Maldjian, J.A., et al.: Automated white matter total lesion volume segmentation in diabetes. Am. J. Neuroradiol. **34**(12), 2265–2270 (2013)

19. Manjón, J., Coupé, P., Martí-Bonmatí, L., Collins, D., Robles, M.: Adaptive non-local means denoising of MR images with spatially varying noise levels. J. Magn. Resonan. Imaging: JMRI **31**(1), 192–203 (2010)

20. Murray, M.E., et al.: A quantitative postmortem MRI design sensitive to white matter hyperintensity differences and their relationship with underlying pathology. J. Neuropathol. Exp. Neurol. **71**(12), 1113–1122 (2012)

21. Oktay, O., et al.: Attention U-net: learning where to look for the pancreas. arXiv preprint arXiv:1804.03999 (2018)

22. Rahil, M., Anoop, B., Girish, G., Kothari, A.R., Koolagudi, S.G., Rajan, J.: A deep ensemble learning-based CNN architecture for multiclass retinal fluid segmentation in oct images. IEEE Access (2023)

23. Rashid, T., et al.: DeepMIR: a deep neural network for differential detection of cerebral microbleeds and iron deposits in MRI. Sci. Rep. **11**(1), 14124 (2021)

24. Ronneberger, O., Fischer, P., Brox, T.: U-net: convolutional networks for biomedical image segmentation. In: Navab, N., Hornegger, J., Wells, W.M., Frangi, A.F. (eds.) MICCAI 2015. LNCS, vol. 9351, pp. 234–241. Springer, Cham (2015). https://doi.org/10.1007/978-3-319-24574-4_28

25. Roseborough, A.D., et al.: Post-mortem 7 tesla MRI detection of white matter hyperintensities: a multidisciplinary voxel-wise comparison of imaging and histological correlates. NeuroImage: Clin. **27**, 102340 (2020)

26. Samaille, T., et al.: Contrast-based fully automatic segmentation of white matter hyperintensities: method and validation. PLoS ONE **7**(11), e48953 (2012)

27. Simões, R., et al.: Automatic segmentation of cerebral white matter hyperintensities using only 3d flair images. Magn. Reson. Imaging **31**(7), 1182–1189 (2013)

28. Smith, C.D., Snowdon, D.A., Wang, H., Markesbery, W.R.: White matter volumes and periventricular white matter hyperintensities in aging and dementia. Neurology **54**(4), 838–842 (2000)

29. Tustison, N.J., et al.: N4ITK: improved n3 bias correction. IEEE Trans. Med. Imaging **29**(6), 1310–1320 (2010)

30. Verhaaren, B.F., et al.: Multiethnic genome-wide association study of cerebral white matter hyperintensities on MRI. Circulat. Cardiovasc. Genet. **8**(2), 398–409 (2015)

31. Viteri, J.A., Loayza, F.R., Pelaez, E., Layedra, F.: Automatic brain white matter hypertinsities segmentation using deep learning techniques. In: HEALTHINF, pp. 244–252 (2021)

32. Zhuang, F.J., Chen, Y., He, W.B., Cai, Z.Y.: Prevalence of white matter hyperintensities increases with age. Neural Regen. Res. **13**(12), 2141 (2018)

Stroke Outcome and Evolution Prediction from CT Brain Using a Spatiotemporal Diffusion Autoencoder

Adam Marcus[1]([✉])(iD), Paul Bentley[1](iD), and Daniel Rueckert[1,2](iD)

[1] Imperial College London, London, UK
{adam.marcus11,p.bentley,d.rueckert}@imperial.ac.uk
[2] Technische Universität München, München, Germany
daniel.rueckert@tum.de

Abstract. Stroke is a major cause of death and disability worldwide. Accurate outcome and evolution prediction has the potential to revolutionize stroke care by individualizing clinical decision-making leading to better outcomes. However, despite a plethora of attempts and the rich data provided by neuroimaging, modelling the ultimate fate of brain tissue remains a challenging task. In this work, we apply recent ideas in the field of diffusion probabilistic models to generate a self-supervised semantically meaningful stroke representation from Computed Tomography (CT) images. We then improve this representation by extending the method to accommodate longitudinal images and the time from stroke onset. The effectiveness of our approach is evaluated on a dataset consisting of 5,824 CT images from 3,573 patients across two medical centers with minimal labels. Comparative experiments show that our method achieves the best performance for predicting next-day severity and functional outcome at discharge.

Keywords: Stroke · Computed Tomography · Diffusion model · Outcome prediction

1 Introduction

Stroke is a major global health problem [25]. It often begins as the result of impaired blood flow in the brain due to a blood clot. As the brain becomes damaged, it swells, which leads to visible changes in Computed Tomography (CT) scans. Accordingly, imaging plays an essential role, and stroke patients frequently have multiple scans throughout their recovery. The impact of a stroke is usually significant and ranges from reduced health-related quality of life to disability and death. Accurately predicting these consequences and how the disease evolves could revolutionize stroke management and usher in a new age of precision medicine. In this era, aftercare decisions could be optimal, and personalized treatment choices, along with patient-specific rehabilitation targets, could lead to better outcomes [6]. Despite numerous attempts, however, realizing this goal remains far.

© The Author(s), under exclusive license to Springer Nature Switzerland AG 2023
A. Abdulkadir et al. (Eds.): MLCN 2023, LNCS 14312, pp. 153–162, 2023.
https://doi.org/10.1007/978-3-031-44858-4_15

Related Work. Current approaches have focused on predicting the functional outcome of treatment and, to a lesser extent, stroke severity. Functional outcome is generally measured using the modified Rankin Scale (mRS) [31], that ranges from 0 (no disability) to 6 (death) and can be dichotomized into independent (<3) or requires assistance (≥ 3). Severity is often assessed with the National Institutes of Health Stroke Scale (NIHSS) [9], that ranges from 0 (no symptoms) to 42 (most severe neurologic deficit). The majority of studies have exclusively used clinical information to predict these outcomes with methods such as logistic regression [16,33], support vector machines [5], random forests [30], and artificial neural networks (ANN) [1,30]. By comparison, fewer studies have attempted to utilize imaging data. Notably, Bacchi et al. [2] proposed using a 3D convolutional neural network (CNN) based on the VGG architecture [29] to process non-contrast CT (NCCT) combined with an ANN. They achieved an area under the receiver operator characteristic curve (AUC) for next-day improvement in severity classification of 0.70. Nawabi et al. [24] applied a random forest model to radiomic features derived from NCCT and attained an AUC for dichotomized mRS at discharge of 0.80. A number of studies have also applied convolutional models to predict outcomes using other modalities, such as magnetic resonance imaging (MRI) [7]. These works have all relied on supervised learning techniques, thereby overlooking the potential benefit of unlabeled images, often more readily available in the medical domain.

The scarcity of labeled images, particularly in medicine, has spurred significant interest self-supervised learning (SSL). In SSL, methods can be broadly categorized into generative and discriminative. Generative methods typically involve an autoencoder that attempts to learn a compressed representation of its input data. Discriminative methods involve solving a task by learning a decision boundary through its data. For vision problems, discriminative techniques are currently considered the more performant. Specifically, joint-embedding approaches, such as contrastive learning [10], that aims to align the embedded representations of augmented views of the same image. However, recent developments enabling diffusion models, which have achieved remarkable performance in generative tasks, to be used as an autoencoder may change this [26].

Contributions. In this work, our objective is to derive an imaging-based feature representation that faithfully captures the entire stroke trajectory from onset to recovery. Our method utilizes diffusion probabilistic models and is motivated by recent developments that allow their use as an autoencoder; the hope is that their higher fidelity image reconstruction translates to a better representation for outcome prediction. The main contributions are: (1) We apply diffusion probabilistic models to generate a self-supervised semantically meaningful stroke representation. (2) We then improve this representation by extending the method to accommodate longitudinal images and the time from stroke onset. (3) We evaluate the effectiveness of these representations when applied to baseline images to predict valuable stroke outcomes, namely, next-day severity and functional status at discharge.

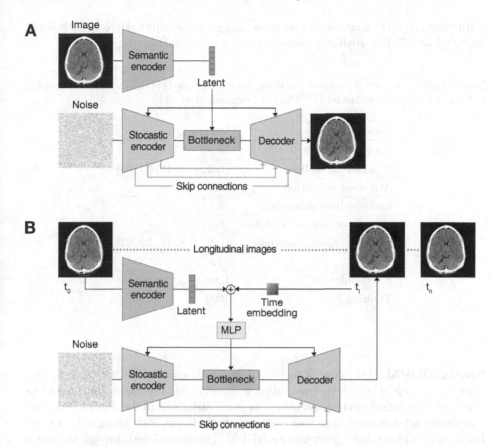

Fig. 1. Overview of the (A) spatial and (B) spatiotemporal diffusion autoencoder approaches. In both cases, a semantic encoder takes an image containing a stroke lesion and maps it to a latent code. A Denoising Diffusion Probabilistic Models (DDPM) [18] is then conditioned on this latent code to denoise a different image of the same lesion taken either at (A) the same time or (B) a future point in time. For the spatiotemporal method, the latent code is also concatenated with the future time using a multilayer perceptron (MLP). After training, the semantic encoder can then be fine-tuned with minimal data and used to predict a stroke outcome.

2   Method

Our spatial method is based on the recently proposed diffusion autoencoder [26], which we then extend to incorporate longitudinal images as our spatiotemporal approach. An overview of these approaches can be seen in Fig. 1. Central to both of our methods is the use of Denoising Diffusion Probabilistic Models (DDPM) [18] that model a distribution of images by learning a denoising process. A successful process can predict the varying amount of Gaussian noise, $\epsilon \sim \mathcal{N}(\mathbf{0}, \mathbf{I})$ with respect to timesteps t (out of T), added to a clean image $\mathbf{x}_0$, by learning a function $\epsilon_\theta(\mathbf{x}_t, t)$ that takes the noisy image $\mathbf{x}_t$ as input. Here $\epsilon_\theta(\cdot)$ is often modeled as a U-Net with parameters θ and a loss function $||\epsilon_\theta(\mathbf{x}_t, t) - \epsilon||$.

Table 1. Network architecture of our diffusion model based on the improved Denoising Diffusion Probabilistic Model (DDPM) of Dhariwal et al. [11].

Parameter	Value
Base channels	16
Channel multipliers	[1, 2, 4, 8, 16, 32, 64]
Attention resolution	[16]
Encoder base channels	16
Encoder attention resolution	[16]
Encoder channel multipliers	[1, 2, 4, 8, 16, 32, 64]
z size	512
β scheduler	Linear
Training T	1000

Spatial DDPM. Our goal is to train a semantic encoder $\text{Enc}(\mathbf{x})$ able to produce a meaningful stroke representation $\mathbf{z}$. Ideally, this representation should be invariant to changes in view and other low-level variations in the image. We therefore design a framework that learns this semantic latent code along with a noisy latent code which results from using a DDPM. This is achieved through the use of a DDPM conditioned on the semantic latent code. In our experiments we employ a ResNet-50 model [14] as the semantic encoder to produce a vector of dimension $d = 512$, which was chosen to resemble the style vector in StyleGAN [19]. During training, a pair of 2D image slices $(\mathbf{x}_a, \mathbf{x}_b)$ showing the same stroke lesion are sampled and each slice is augmented with a CT-specific strategy including: random axial plane flips; $\pm 5\%$ isotropic scaling; $\pm 20\,\text{mm}$ translation; and $\pm 0.5\,\text{rad}$ axial rotation. These images are then used as inputs to the semantic encoder and DDPM which are jointly optimized with a revised $\mathcal{L}_{simple}$ loss [18]:

$$z = \text{Enc}(\mathbf{x}_a)$$

$$\mathcal{L}_{simple} = \sum_{t=1}^{T} \mathbb{E}_{\mathbf{x}_a, \mathbf{x}_b, \epsilon_t} \left[||\epsilon_\theta(\mathbf{x}_{b_t}, t, z) - \epsilon_t||_2^2 \right] \tag{1}$$

Here $\epsilon_t \in \mathbb{R}^{h \times w} \sim \mathcal{N}(\mathbf{0}, \mathbf{I})$, $h \times w$ is the spatial resolution, $\mathbf{x}_{b_t} = \sqrt{\alpha_t}\mathbf{x}_b + \sqrt{1 - \alpha_t}\epsilon_t$, $T = 1000$ or an equally large number, and α relates to the variance schedule of the Guassian noise [18]. Our conditional DDPM is implemented as a modified U-Net [27] with the improvements by Dhariwal et al. [11] that include increasing the attention channels and using BigGAN [8] residual blocks for up and downsampling. We make two additional architectural changes with the other hyperparameters specified in Table 1. First, we introduce group normalization (GroupNorm) [34] layers after every 1×1 convolution, acting as skip connections in each residual block. Empirically we found this stabilizes training and is particularly effective when using a deeper network with more blocks. Second, we replace the other GroupNorm layers throughout the network with Adaptive Spatial Group Normalization (AdaSpaGN) that scales and shifts the normalized feature maps. This is defined where $k \in \mathbb{R}^{c\times h\times w}$ is the feature maps with channels c obtained from the U-Net, diffusion timesteps is t, the semantic latent code is z, ψ is a sinusoidal encoding function [32] and MLP is a multilayered perceptron:

$$(s,b) \in \mathbb{R}^{2\times c} = \mathrm{MLP}(z, \psi(t))$$
$$\mathrm{AdaSpaGN}(k, s, b) = s\mathrm{GroupNorm}(k) + b \tag{2}$$

Spatiotemporal DDPM. To make the representation produced by the semantic encoder more suited for clinical outcome prediction, our key insight is that it should be invariant to the point in time that the image was taken along the stroke trajectory. We, therefore, extend our spatial method by sampling a different pair of images during training showing the same lesion but now at distinct points in time. The semantic encoder receives $\mathbf{x}_a$, which is always a view from the earliest scan, and the DDPM receives a noisy variant of $\mathbf{x}_b$ taken at a later time. In our ablation study, we explore whether it is necessary to enforce this ordering.

Intending to reduce the difficulty of learning this new task, we also condition the DDPM on the time image $\mathbf{x}_b$ was taken since stroke onset. We log-transform these times to account for a skewed distribution and replace the AdaSpaGN layers with Adaptive Temporal Group Normalization (AdaTempGN) which is defined where time is n as:

$$(s,b) \in \mathbb{R}^{2\times c} = \mathrm{MLP}(z, \psi(\log(n)), \psi(t))$$
$$\mathrm{AdaTempGN}(k, s, b) = s\mathrm{GroupNorm}(k) + b \tag{3}$$

3 Experiments

3.1 Materials

Dataset. Experiments were performed on a dataset of 3,573 acute ischemic stroke patients collected across two clinical sites from 2010 to 2019. Patients were divided so that a fixed random 20% split were used for testing and the remainder for training and validation using a five-fold cross-validation approach. Table 2 details the characteristics of these groups. A high number of patients were missing outcome measures but contained timing information. Muschelli's [23] recommended pipeline was used to extract and anonymize the patients' non-contrast CT images. Full ethical approval was granted by Wales REC 3 reference number 16/WA/0361.

Table 2. Population characteristics of the clinical dataset.

Characteristic	Train and validation set (n = 2858)	Test set (n = 715)
Number of CT, mean (range)	1.63 (1–5)	1.63 (1–5)
Missing outcome, n (%)	1178 (41.2%)	276 (38.6%)
Age (years), median (IQR)	75.0 (62.4–83.7)	75.0 (65.0–83.0)
Female sex, n (%)	914 (54.4%)	242 (55.1%)
ASPECTS, median (IQR)	10 (9–10)	10 (9–10)
NIHSS on admission, median (IQR)	7 (4–13)	7 (4–13)
NIHSS at 24 h, median (IQR)	5 (2–11)	5 (2–12)
mRS on discharge, median (IQR)	3 (1–4)	3 (1–4)
Time from symptom onset to first CT (minutes), median (IQR)	180 (95–522)	169 (90–550)

IQR = Interquartile range; ASPECTS = Alberta stroke program early CT score; NIHSS = National Institutes of Health Stroke Scale [9]; mRS = Modified Rankin score [31]

Evaluation. A positive stroke outcome was defined as either a mRS <3 or the improvement in next-day NIHSS of ≥ 4 points. These thresholds were selected due to their use in previous studies [2,13,28]. Classification performance was evaluated using accuracy (ACC), F1 score, and AUC. To determine significant differences in AUC, permutation testing was used [3]. The Fréchet inception distance (FID) [17] and mean squared error (MSE) were used to assess the quality of image reconstruction.

Implementation. All models were implemented using PyTorch version 1.13 on a machine with 3.80 GHz Intel® Core™ i7-10700K CPU and an NVIDIA GeForce RTX 3080 10 GB GPU. The AdamW [21] optimizer was used with a learning rate of 10^{-3} when training from scratch and 10^{-4} when fine tuning with a weight decay coefficient of 10^{-2}. To ensure fairness, all models were trained initially for 1 million steps and a ResNet-50 [14] was used as the semantic encoder. If required, the encoder was then independently fine-tuned for each prediction task, for an additional 100,000 steps, with all but the final layer weights frozen for the first 10,000 steps. Lesion containing slices were identified using a previously described model [22] and linearly sampled from the original volumes to a uniform size of 512×512 with a spatial resolution of $0.4 \times 0.45\,\text{mm}^2$. Images were clipped based on the 0.5 and 99.5th percentile then normalized using Z-score. Final inference of the models required approximately a second per subject.

3.2 Results

Comparison with Baseline. The quantitative results can be seen in Table 3. First, we recognize that across all the models tested, predicting improvement in severity over 24 h appears to be a more challenging task than functional status at discharge. It seems plausible that this may be due to the models not knowing the given treatment, which is likely more impactful in the short term. Second, both our method and autoencoders appear to increase performance over

Table 3. Stroke outcome prediction results obtained by our method and spatiotemporal ablation variants compared to baseline approaches. In our method, a pair of 2D image slices, $\mathbf{x}_a$ and $\mathbf{x}_b$, showing the same stroke lesion are sampled, where $\mathbf{x}_a$ is always from the earliest scan and the DDPM receives a noisy variant of $\mathbf{x}_b$ taken at a later time. In the *Any forward pair* variant, $\mathbf{x}_a$ is from any time prior to $\mathbf{x}_b$, and in the *Any pair* variant, $\mathbf{x}_a$ and $\mathbf{x}_b$ can be views from any time. For fair comparison all methods utilized a ResNet-50 [14] as the semantic encoder and followed an identical fine-tuning procedure, unless trained directly.

Method		24-h NIHSS			Discharge mRS			FID	MSE
		AUC	ACC	F1	AUC	ACC	F1		
CNN	Direct training	0.584	59.9	52.0	0.702	60.4	57.4	—	—
	VICReg [4]	0.582	59.9	52.0	0.711	63.1	58.6	—	—
Autoencoder	Variational [20]	0.628	62.9	63.6	0.726	64.4	63.5	8.9	69.9
	Diffusion [26]	0.623	62.6	64.2	0.735	65.8	64.2	**7.1**	**47.7**
Ours	Spatial	0.648	63.6	65.4	0.757	67.7	69.2	7.3	47.9
	Spatiotemporal	**0.669**	63.6	67.4	0.788	**70.8**	71.6	7.4	48.2
	No augmentation	0.666	63.6	67.4	0.785	70.2	69.8	7.4	47.9
	Any forward pair	0.663	63.3	**67.6**	**0.789**	70.6	71.1	7.3	48.1
	Any pair	0.667	**63.8**	67.2	0.787	70.8	**72.0**	7.3	48.2

NIHSS = National Institutes of Health Stroke Scale [9]; mRS = Modified Rankin score [31]; AUC = Area under the receiver operator characteristic curve; ACC = Accuracy; FID = Fréchet inception distance [17]; MSE = Mean squared error; CNN = Convolutional Neural Network

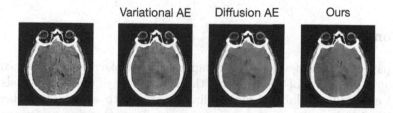

Fig. 2. Example reconstructed image of a right middle cerebral artery (MCA) stroke from our test set for different methods. The performance of our spatiotemporal approach and the diffusion autoencoder (AE) are similar and superior to a variational AE.

training directly from the labeled images. This is also true when employing non-contrastive pre-training with VICReg [4], which was selected over other methods due to its robustness to smaller batch sizes. Third, we note that the diffusion-based models provide superior image reconstruction ability, which is further supported by qualitative evaluation seen in Fig. 2. Fourth, despite offering the best prediction performance, our approach slightly underperforms in image reconstruction. This suggests that subtle local image features, which global image metrics may not fully capture, hold greater significance in outcome prediction. Finally, we observe that our spatiotemporal approach shows significantly (p value ≤ 0.05) greater AUC for predicting discharge mRS over other methods.

Comparison with the Literature. There are few existing studies to which we can fairly compare our results. Partly this is due to most studies focusing on predicting the justifiably more clinically relevant 90-day rather than discharge mRS. This is a limitation of the current work, as the dataset used did not contain longer-term measures. An issue further magnified by the lack of a readily available public dataset. However, it should be noted that both next-day NIHSS and discharge mRS are highly associated with 90-day mRS [12,15]. In comparison with Bacchi et al. [2], who also predicted an improvement in 24-h NIHSS, we achieved a higher AUC of 0.669 compared to 0.63 using only imaging features. Similarly, we attained a comparable AUC for predicting dichotomized discharge mRS of 0.789 to 0.80 by Nawabi et al. [24]. Although noteworthy that they looked at hemorrhagic rather than ischemic strokes.

Ablation Study. To justify our design decisions and verify the effectiveness of our approach, we conducted several ablation experiments, shown in Table 3. We first note that our augmentation strategy has a minimal impact on the final stroke representation leading to a similar performance in outcome prediction. Similarly, we observed no significant effect when changing the combination of images used during training. We hypothesize this may be partly due to our dataset often containing only a single pair of images per subject.

4   Conclusion

In this paper, we have developed an imaging-based stroke representation capturing features predictive of future events through the use of diffusion models. Our method can utilize unlabeled longitudinal images with any duration between scans. Empirical results suggest our approach offers promising performance at estimating next-day severity and functional status at discharge. Future research includes integrating clinical information and prospective validation with longer-term outcomes. We hope that this work ultimately leads to more effective personalized treatment strategies for stroke patients.

Acknowledgements. This work was supported by the UK Research and Innovation: UKRI Center for Doctoral Training in AI for Healthcare under Grant EP/S023283/1 and UK National Institute for Health Research i4i Program under Grant II-LA-0814–20007.

References

1. Asadi, H., Dowling, R., Yan, B., Mitchell, P.: Machine learning for outcome prediction of acute ischemic stroke post intra-arterial therapy. PloS One **9**(2), e88225 (2014)
2. Bacchi, S., Zerner, T., Oakden-Rayner, L., Kleinig, T., Patel, S., Jannes, J.: Deep learning in the prediction of ischaemic stroke thrombolysis functional outcomes: a pilot study. Acad. Radiol. **27**(2), e19–e23 (2020)
3. Bandos, A.I., Rockette, H.E., Gur, D.: A permutation test sensitive to differences in areas for comparing roc curves from a paired design. Stat. Med. **24**(18), 2873–2893 (2005)
4. Bardes, A., Ponce, J., LeCun, Y.: Vicreg: variance-invariance-covariance regularization for self-supervised learning. arXiv preprint arXiv:2105.04906 (2021)
5. Bentley, P., et al.: Prediction of stroke thrombolysis outcome using ct brain machine learning. NeuroImage: Clin. **4**, 635–640 (2014)
6. Bonkhoff, A.K., Grefkes, C.: Precision medicine in stroke: towards personalized outcome predictions using artificial intelligence. Brain **145**(2), 457–475 (2022)
7. Bourached, A., et al.: Scaling behaviors of deep learning and linear algorithms for the prediction of stroke severity. In: medRxiv, pp. 2022–12 (2022)
8. Brock, A., Donahue, J., Simonyan, K.: Large scale gan training for high fidelity natural image synthesis. arXiv preprint arXiv:1809.11096 (2018)
9. Brott, T., et al.: Measurements of acute cerebral infarction: a clinical examination scale. Stroke **20**(7), 864–870 (1989)
10. Chen, T., Kornblith, S., Norouzi, M., Hinton, G.: A simple framework for contrastive learning of visual representations. In: International Conference on Machine Learning, pp. 1597–1607. PMLR (2020)
11. Dhariwal, P., Nichol, A.: Diffusion models beat gans on image synthesis. Adv. Neural Inf. Process. Syst. **34**, 8780–8794 (2021)
12. ElHabr, A.K., et al.: Predicting 90-day modified rankin scale score with discharge information in acute ischaemic stroke patients following treatment. BMJ Neurol. Open **3**(1) (2021)
13. Hacke, W., et al.: Thrombolysis with alteplase 3 to 4.5 h after acute ischemic stroke. New Engl. J. Med. **359**(13), 1317–1329 (2008)
14. He, K., Zhang, X., Ren, S., Sun, J.: Deep residual learning for image recognition. In: Proceedings of the IEEE Conference on Computer Vision and Pattern Recognition, pp. 770–778 (2016)
15. Hendrix, P., et al.: Nihss 24 h after mechanical thrombectomy predicts 90-day functional outcome. Clin. Neuroradiol. **32**(2), 401–406 (2022)
16. Heo, J., Yoon, J., Park, H.J., Kim, Y.D., Nam, H.S., Heo, J.H.: Machine learning-based model can predict stroke outcome. Stroke **49**(Suppl_1), A194–A194 (2018)
17. Heusel, M., Ramsauer, H., Unterthiner, T., Nessler, B., Hochreiter, S.: Gans trained by a two time-scale update rule converge to a local nash equilibrium. Adv. Neural Inf. Process. Syst. **30** (2017)
18. Ho, J., Jain, A., Abbeel, P.: Denoising diffusion probabilistic models. Adv. Neural Inf. Process. Syst. **33**, 6840–6851 (2020)

19. Karras, T., Laine, S., Aila, T.: A style-based generator architecture for generative adversarial networks. In: Proceedings of the IEEE/CVF Conference on Computer Vision and Pattern Recognition, pp. 4401–4410 (2019)
20. Kingma, D.P., Welling, M.: Auto-encoding variational bayes. arXiv preprint arXiv:1312.6114 (2013)
21. Loshchilov, I., Hutter, F.: Decoupled weight decay regularization. arXiv preprint arXiv:1711.05101 (2017)
22. Marcus, A., Bentley, P., Rueckert, D.: Concurrent ischemic lesion age estimation and segmentation of ct brain using a transformer-based network. In: Machine Learning in Clinical Neuroimaging: 5th International Workshop, MLCN 2022, Held in Conjunction with MICCAI 2022, Singapore, 18 September 2022, Proceedings, pp. 52–62. Springer, Heidelberg (2022). DOI: https://doi.org/10.1007/978-3-031-17899-3_6
23. Muschelli, J.: Recommendations for processing head ct data. Front. Neuroinf. **13**, 61 (2019)
24. Nawabi, J., et al.: Imaging-based outcome prediction of acute intracerebral hemorrhage. Transl. Stroke Res. **12**, 958–967 (2021)
25. Organization, W.H.: Global health estimates (2018). https://www.who.int/healthinfo/global_burden_disease/en/
26. Preechakul, K., Chatthee, N., Wizadwongsa, S., Suwajanakorn, S.: Diffusion autoencoders: toward a meaningful and decodable representation. In: Proceedings of the IEEE/CVF Conference on Computer Vision and Pattern Recognition, pp. 10619–10629 (2022)
27. Salimans, T., Goodfellow, I., Zaremba, W., Cheung, V., Radford, A., Chen, X.: Improved techniques for training gans. Adv. Neural Inf. Process. Syst. **29** (2016)
28. Samak, Z.A., Clatworthy, P., Mirmehdi, M.: Fema: feature matching auto-encoder for predicting ischaemic stroke evolution and treatment outcome. Computer. Med. Imaging Graph. **99**, 102089 (2022)
29. Simonyan, K., Zisserman, A.: Very deep convolutional networks for large-scale image recognition. arXiv preprint arXiv:1409.1556 (2014)
30. Van Os, H.J., et al.: Predicting outcome of endovascular treatment for acute ischemic stroke: potential value of machine learning algorithms. Front. Neurol. **9**, 784 (2018)
31. Van Swieten, J., Koudstaal, P., Visser, M., Schouten, H., Van Gijn, J.: Interobserver agreement for the assessment of handicap in stroke patients. Stroke **19**(5), 604–607 (1988)
32. Vaswani, A., et al.: Attention is all you need. Adv. Neural Inf. Process. Syst. **30** (2017)
33. Venema, E., et al.: Selection of patients for intra-arterial treatment for acute ischaemic stroke: development and validation of a clinical decision tool in two randomised trials. BMJ **357** (2017)
34. Wu, Y., He, K.: Group normalization. In: Proceedings of the European Conference on Computer Vision (ECCV), pp. 3–19 (2018)

Morphological Versus Functional Network Organization: A Comparison Between Structural Covariance Networks and Probabilistic Functional Modes

Petra Lenzini[1]([✉]), Tom Earnest[1], Sung Min Ha[1], Abdalla Bani[1], Aristeidis Sotiras[1,2], and Janine Bijsterbosch[1]

[1] Department of Radiology, Washington University in St. Louis, St. Louis, USA
plenzini@wustl.edu

[2] Institute for Informatics, Data Science, and Biostatistics, Washington University in St. Louis, St. Louis, USA

Abstract. The degree to which gray matter morphology constrains brain func tion remains an elusive target of investigation due to the lack of a gold-standard against which to argue for a better or worse metric of neurobiological significance. Therefore, we sought to compare the output of state-of-the-art morphological and functional covariance decomposition methods directly to one another. Specifically, we compared the spatial network organization produced by non-negative matrix factorization of T1-weighted images and probabilistic functional modes of resting state functional MRI scans from 1297 UK Biobank subjects. We measured the cosine similarity of matched networks across 2 to 140 rank decompositions. Our findings revealed strong commonality between morphological and functional networks at the lowest rank (2). Morphology-function network commonality was retained across all ranks in the visual cortex, but broader network organization diverged between morphology and function at higher ranks.

Keywords: NMF · probabilistic functional modes · T1 · rsfMRI

1 Introduction

Prior work has provided robust evidence for the presence of a multivariate network organization in functional neuroimaging data (particularly resting state) using a variety of data-driven decomposition methods [1–5]. In recent years, data decomposition techniques have been used to identify covariance networks from T1-weighted MRI data, too [6–8]. While the association between functional connectivity and white matter connectivity measured with diffusion weighted imaging has been investigated extensively [9], the link between brain morphology measured with T1-weighted imaging and functional network organization is poorly understood; it is unclear to what degree the whole-brain network organization converges across morphology and function.

A. Sotiras and J. Bijsterbosch—Shared senior author.

© The Author(s), under exclusive license to Springer Nature Switzerland AG 2023
A. Abdulkadir et al. (Eds.): MLCN 2023, LNCS 14312, pp. 163–172, 2023.
https://doi.org/10.1007/978-3-031-44858-4_16

Prior work has shown similar associations with behavior between morphology and function [10], suggesting the presence of shared individual differences. Prior work has also identified similarities in organizational gradients along a sensorimotor-association cortex axis derived from structural and functional imaging data [11], and has shown convergent morphological and functional patterns within specific brain regions [12–14]. Moreover, morphological networks have been shown to predict functional connectivity patterns [6, 15]. As such, we hypothesized that the spatial organization of networks derived from T1 data would be similar to those derived on resting state functional data, especially in (but not limited to), the primary sensorimotor cortex.

In order to compare morphological and functional network organization, we calculated networks using state-of-the-art approaches. We performed non-negative matrix factorization (NMF) [8, 16] of T1-weighted images restricted to gray matter volumes. NMF factorizes the collection of vectorized images into two non-negative matrices representing the spatial layout of morphological networks and the subject-specific weights for each network. Due to the non-negativity, projectivity, and orthogonality restraints, images are the sum of these weighted, non-overlapping parts and residual noise. NMF morphological network organization has previously been linked to development [15] and psychopathology [17]. For resting state fMRI (rsfMRI) images, we computed probabilistic functional modes (PROFUMO) [18]. PROFUMO estimates Bayesian hierarchical models on both spatial topographies and functional connectivity, simultaneously, leveraging knowledge about the hemodynamic response function for temporal relationships in space, as well as a delta-Gaussian distribution for detection of non-zero signal intensity above noise. PROFUMO functional network organization has previously been linked to individual differences in behavior [27, 28].

2 Methods

2.1 Subjects

MRI and phenotypic data were downloaded from the UK Biobank Resource (UKB) with the "Cross-diagnostic and cross-platform multimodal analysis of UK Biobank imaging data" application (id: 47267). A subsample of UKB participants were selected to avoid psychopathology that may alter morphological and/or functional network organization, and to ensure maximum data quality. Specifically, healthy UKB subjects were defined as subjects having no ICD-10 diagnoses at baseline (variable id: 41270). Healthy subjects with T1-weighted structural and rsfMRI scans at instance 2 (first imaging session) were included if they had non-missing intracranial volume, age, sex, site, and mean resting state or task head motion. Individuals with resting state head motion in the lowest 25th percentile were selected from this group, resulting in a final sample size of 1297 subjects (43% male, mean age 60 ± 6.9 years).

2.2 Image Preprocessing

T1-weighted scans (1 mm isotropic voxels, TR = 2000 ms, TI = 880 ms) were preprocessed and segmented with FAST [19] as described previously and released through the

UKB showcase [20]. Gray matter (GM) probability maps were transformed to 1 mm MNI space using linear and nonlinear transforms derived from FNIRT (transformation matrices released). GM-probability maps in MNI 1 mm space were modulated by the determinant of the Jacobian of their MNI registration warps, then down-sampled to 2 mm using the nearest neighbor algorithm with FLIRT. With the masks described below, subject-specific smoothing of modulated and down-sampled images proceeded in two steps to avoid signal degradation at mask boundaries as described in [3]. Briefly: after images were masked, they were smoothed (sigma = 2 in fslmaths) and masked again. Then, images were divided by the smoothed mask (again, sigma = 2) to attenuate smoothing at mask boundaries. Resting state fMRI scans (2.4 mm isotropic voxels, TR = 735 ms, TE = 39 ms, multiband factor 8) were preprocessed as described previously [20], and transformed into 2 mm MNI space using the linear and non-linear transforms provided. The resting state data in 2 mm MNI space was then smoothed and masked like the GM probability maps using the masks derived on the GM probability maps.

2.3 Masks

Two different masks were used in different parts of this work, namely a whole-brain GM mask and a cortical-only GM mask. In order to create these group masks, the structural GM probability maps in MNI space for each subject were down-sampled to 2 mm, thresholded at 0.5 and averaged across all 1297 participants. The resulting group average GM mask was thresholded at 0.4, constituting the whole-brain GM mask. For the cortical-only GM mask, we then found the intersection of our previously generated GM mask, and the union of the 7 down-sampled networks defined by [5]. This cortical-only GM mask contained no cerebellum or subcortical regions.

2.4 Models

Orthonormal projective non-negative matrix factorization (NMF) with non-negative double singular value decomposition (NNDSVD) [16, 21, 22] was performed in MATLAB on pre-processed GM density maps. Probabilistic functional modes were generated on pre-processed resting state scans, with hemodynamic response function (HRF) modeling turned on and repetition time (TR) = 0.735 s, using the default HRF. For reference, PROFUMO and NMF were compared with independent components (ICA) generated using MELODIC, with TR = .735 s. All methods require as input a user-specified parameter determining the rank of the estimated decomposition. Accordingly, several runs of NMF and PROFUMO (and ICA, where applicable) were performed to enable cross-method comparisons.

For the analyses described below, group modes from NMF, PROFUMO and ICA were compared with cosine similarity (CS) and matched using a modified Jonker-Volgenant algorithm (optimize.linear_sum_assignment; [24]). To match the non-negative constraint in NMF, negative values in PROFUMO and ICA modes were set to zero before calculating CS. Network similarity was assessed with the proportion of matched components having CS > 0.5 and adjusted rand index (ARI) with a 'winner-takes-all' approach in PROFUMO. Code for this paper is available at: https://github.com/petralenzini/NMF_Profumo_UKB.

2.5 Assessing Decomposition Stability Across Ranks

We first sought to understand whether the stability of NMF and PROFUMO network decompositions varies as a function of rank, using split-half analysis. Specifically, we used the 'Anticlust' R-package [23] to generate two sample splits (N = 649 and N = 648, respectively) that were matched for site, age, sex, intracranial volume, and mean head motion during resting state and task scan acquisitions. Structural and functional data from each split were masked using the whole-brain GM mask and given as input to the respective methods. Solutions were estimated for a wide range of ranks (i.e., 2, 4, 6, 8, 10, 12, 14, 16, 18, 20, 22, 24, 26, 28, 30, 32, 36, 40, 50, 60, 80, 100, 120, 140).

2.6 Assessing the Impact of NMF Orthogonality Constraints

NMF and PROFUMO impose different constraints in the derived data decompositions. Importantly, NMF imposes an orthogonality constraint, which results in non-overlapping components. In contrast, there is no such constraint in PROFUMO, which often results in overlapping components. Accordingly, we aimed to assess the impact of the differences in orthogonality constraints on morphology-function comparisons. To this end, we included ICA in our comparisons, which tends to produce spatially non-overlapping components. Input data were masked separately with both the whole-brain GM mask and cortical-only GM mask and ICA, NMF, and PROFUMO decompositions were derived for ranks 2 and 25.

2.7 Comparing NMF and PROFUMO Decompositions Across Ranks

We examined whether convergence between structure and function varies as a function of the decomposition rank. Accordingly, we compared NMF and PROFUMO solutions across the following ranks: 10, 20, 40, 60, 80, 100, 120, 140. All input data were masked using the whole-brain GM mask.

2.8 Comparison of Individual Differences Scores

PROFUMO produces several subject-specific outputs, including network maps, network time series, and network amplitudes. NMF, however, includes only one type of subject-specific output, namely non-negative network weights. We tested the relationship between subject-specific NMF weights and PROFUMO amplitudes in terms of Pearson correlations for the rank 25 decompositions with GM-mask.

3 Results

3.1 Stability of Morphological and Functional Networks

Both PROFUMO and NMF achieved relatively high split-half stability suggesting that morphological and functional networks can be reliability estimated (Fig. 1). Cosine similarity (CS) was greatest at rank 2 for both methods. Mean similarity with PROFUMO decreased with rank. NMF dipped below 0.7 between ranks 12 and 20 before returning to hover around a mean CS of 0.7 by rank 22.

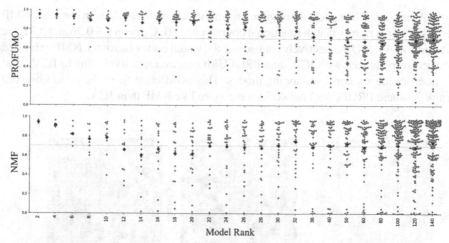

Fig. 1. Split-sample cosine similarity distributions for matched components at various ranks. Note that axes are not linear.

3.2 Similarity Between Morphological and Functional Networks

Primary Sensorimotor and Association Cortex Networks at Low Rank. At low rank (i.e., 2), there was strong correspondence between morphological networks estimated using NMF and functional networks estimated using PROFUMO and ICA (Fig. 2). When masked by the cortical-only GM mask, low-rank networks differentiated between primary sensorimotor cortices and association cortices. CS was higher in the sensory motor cortices (0.75 and 0.49 for PROFUMO and ICA, respectively) than in the association cortices (0.55, 0.43). In contrast, using the more liberal GM-mask resulted in NMF first differentiating between cerebellar and non-cerebellar GM.

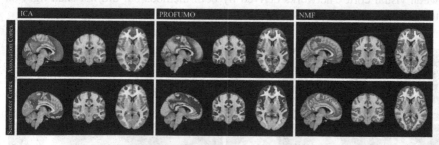

Fig. 2. ICA (green), PROFUMO (blue), and NMF (red) showing similarity at rank 2 decomposition into association cortex (top row) and primary sensorimotor cortex (bottom row). (Color fig online)

Medium Rank Network Similarity and Orthogonality Constraints. Figure 3 shows the pairwise similarity between components produced by the labeled methods. Columns have been reordered in each panel to highlight best-matched network pairs (pairs on the diagonal). As expected, ICA and PROFUMO demonstrated strong 1:1 correspondence

(Fig. 3A, mean CS of matched components on the diagonal = 0.50). In contrast, NMF networks did not tightly match functional counterparts (B, C, mean CS 0.36 and 0.31, for ICA and PROFUMO, respectively) outside of the visual cortex (arrows). NMF exhibited better correspondence with ICA than PROFUMO on average, likely due to ICA's cost function benefitting non-overlapping modes. This is different from the trend observed at rank 2, where PROFUMO networks were more like NMF than ICA.

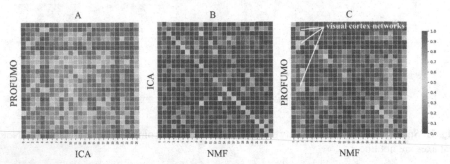

Fig. 3. ICA and PROFUMO have strong 1:1 network correspondence (A). In contrast, NMF networks do not tightly match counterparts (B, C) outside of the visual cortex (arrows).

Differential Breakdown of Morphological and Functional Networks. Morphological and functional networks did not have a clear 1:1 relationship beyond rank 2 when restricted to cortical GM (Fig. 2). Figure 4 illustrates the differential breakdown of the visual cortex at rank 20, where 3 PROFUMO networks were associated with one NMF network. This one-to-many relationship between NMF and PROFUMO for the visual cortex is also highlighted in Fig. 3C (arrows) and was consistently observed across ranks. Hierarchical clustering of network associations recapitulated a strong link between visual cortex networks derived by both methods but did not shed light on a pattern of association above rank 2.

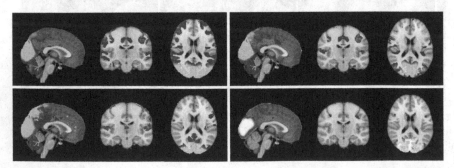

Fig. 4. Spatial arrangement of visual cortex networks for NMF (hot) and PROFUMO (blue) from rank 20 decompositions (GM mask), illustrating a 1-to-many relationship between networks derived by these methods. (Color figure online)

3.3 Similarity Between Networks Across Ranks

The results above suggest a close correspondence between NMF and PROFUMO at low rank, which decreases at medium rank partly due to differential network breakdown (e.g., 3 PROFUMO visual networks mapping onto 1 NMF network). These findings may suggest either a general (linear) trend from convergence towards divergence as a function of rank, or a maximal convergence at differential rank orders (e.g., lower rank PROFUMO mapping onto higher rank NMF). To test these possibilities, we comprehensively compared NMF and PROFUMO as a function of rank (Fig. 5). The results revealed poor convergence even at relatively low rank (10), which does not support the presence of a gradual linear trend from convergence to divergence. Different indices of comparisons suggest different optimal rank comparisons (e.g., PROFUMO 140 - NMF 10 for CS versus PROFUMO 140 - NMF 60 for ARI), suggesting general instability in network mapping between morphology and function with these methods.

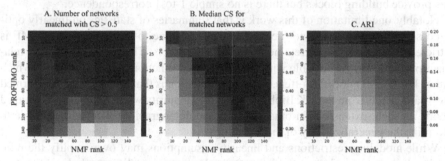

Fig. 5. NMF and PROFUMO network overlap diagnostics by rank (GM mask). A). Number of matched components with cosine similarity (CS) > 0.5. B). Median CS for matched components. C). Adjusted Rand Index after 'Winner Takes All' assignment for overlapping PROFUMO modes. Lighter colors indicate higher convergence.

3.4 Comparison of Individual Differences Scores

Consistent with group level spatial results, subject-specific weights/amplitudes from NMF vs PROFUMO networks displayed minimal correlation (mean r = 0.055 ± 0.0373).

4 Discussion

We compared structural and functional network organization using non-negative matrix factorization (NMF) and probabilistic functional modes (PROFUMO). Although we observed strong and robust similarity of visual cortex networks between morphology and function, agreement between broader network organization was low at all ranks above 2. The network agreement at low rank is consistent with other work demonstrating comparable sensorimotor-association cortical hierarchies observed in structural and functional data [29]. Despite highly stable network decompositions within PROFUMO and NMF (Fig. 1), our results indicate consistently poor network convergence between functional and morphological networks across all ranks above 2.

Our findings revealed strong convergence of visual network organization between morphological and functional decompositions at varying ranks. A possible explanation for these findings is that synchronized functional demands might induce use-dependent coordinated growth [25]. Indeed, if morphology-function network similarities arise as a result of use-dependent coordinated neurodevelopment, this could explain why stronger network similarities were observed in primary sensorimotor networks compared to higher order association networks which experience fewer synchronized functional demands during early childhood development.

Recent work showed that resting state data reconstructed from geometric basis maps derived from morphological data combined with a wave model was relatively highly correlated with empirically measured resting state data [26]. These prior findings may be considered somewhat inconsistent with our results. However, the study by Pang et al. allowed for linear combinations of morphological basis maps to recapitulate function. Taken together, these findings suggest a more complex relationship in which morphology does provide building blocks but there is no simple 1-to-1 correspondence.

Notably, one limitation of this work is that summaries of statistics on linearly optimized sets of networks are not designed to identify one-to-many networks. NMF is restricted (by design) to discrete non-overlapping parcels. Therefore, one-to-one relationship would only be possible with an orthonormal functional decomposition (which neither PROFUMO nor ICA offer). A further limitation is that the study was performed using participants from the UK Biobank, which is a majority white middle-to-older aged cohort. As such, extensions in other datasets and demographics would be valuable to determine the generalizability of these findings.

While modeling restrictions and linearity assumptions may oversimplify the relationship between morphology and function, characterizing the morphology-function relationship in terms of modeling choices remains critical in leveraging the strengths of diverging bodies of research in reaching useful and neurobiologically meaningful consensus on the relationship between morphology and function. We have provided evidence that this relationship is non-linear, and that PROFUMO, ICA, and NMF agree at extremely low dimensionality.

Acknowledgments. This project was funded by the McDonnell Center for Systems Neuroscience at Washington University in St. Louis.

References

1. Beckmann, C.F., DeLuca, M., Devlin, J.T., Smith, S.M.: Investigations into resting-state connectivity using independent component analysis. Philos. Trans. R. Soc. B Biol. Sci. **360**, 1001–1013 (2005). https://doi.org/10.1098/rstb.2005.1634
2. Bielczyk, N.Z., et al.: Thresholding functional connectomes by means of mixture modeling. Neuroimage **171**, 402–414 (2018). https://doi.org/10.1016/j.neuroimage.2018.01.003
3. Farahibozorg, S.-R., et al.: Hierarchical modelling of functional brain networks in population and individuals from big fMRI data (2021). https://doi.org/10.1101/2021.02.01.428496
4. Power, J.D., et al.: Functional network organization of the human brain. Neuron **72**, 665–678 (2011). https://doi.org/10.1016/j.neuron.2011.09.006

5. Yeo, B.T.T., et al.: The organization of the human cerebral cortex estimated by intrinsic functional connectivity. J. Neurophysiol. **106**, 1125–1165 (2011). https://doi.org/10.1152/jn.00338.2011
6. Alexander-Bloch, A., Raznahan, A., Bullmore, E., Giedd, J.: The convergence of maturational change and structural covariance in human cortical networks. J. Neurosci. **33**, 2889–2899 (2013). https://doi.org/10.1523/JNEUROSCI.3554-12.2013
7. Carmon, J., et al.: Reliability and comparability of human brain structural covariance networks. Neuroimage **220**, 117104 (2020). https://doi.org/10.1016/j.neuroimage.2020.117104
8. Sotiras, A., Resnick, S.M., Davatzikos, C.: Finding imaging patterns of structural covariance via Non-Negative Matrix Factorization. Neuroimage **108**, 1–16 (2015). https://doi.org/10.1016/j.neuroimage.2014.11.045
9. Suárez, L.E., Markello, R.D., Betzel, R.F., Misic, B.: Linking structure and function in macroscale brain networks. Trends Cogn. Sci. **24**, 302–315 (2020). https://doi.org/10.1016/j.tics.2020.01.008
10. Llera, A., Wolfers, T., Mulders, P., Beckmann, C.F.: Inter-individual differences in human brain structure and morphology link to variation in demographics and behavior. eLife **8**, e44443 (2019). https://doi.org/10.7554/eLife.44443
11. Huntenburg, J.M., Bazin, P.-L., Margulies, D.S.: Large-scale gradients in human cortical organization. Trends Cogn. Sci. **22**, 21–31 (2018). https://doi.org/10.1016/j.tics.2017.11.002
12. Kelly, C., et al.: A convergent functional architecture of the insula emerges across imaging modalities. Neuroimage **61**, 1129–1142 (2012). https://doi.org/10.1016/j.neuroimage.2012.03.021
13. Zhang, Z., et al.: Resting-state brain organization revealed by functional covariance networks. PLoS ONE **6**, e28817 (2011). https://doi.org/10.1371/journal.pone.0028817
14. Segall, J.M., et al.: Correspondence between structure and function in the human brain at rest. Front. Neuroinform. **6**, 10 (2012). https://doi.org/10.3389/fninf.2012.00010
15. Sotiras, A., Toledo, J.B., Gur, R.E., Gur, R.C., Satterthwaite, T.D., Davatzikos, C.: Patterns of coordinated cortical remodeling during adolescence and their associations with functional specialization and evolutionary expansion. Proc. Natl. Acad. Sci. **114**, 3527–3532 (2017). https://doi.org/10.1073/pnas.1620928114
16. Yang, Z., Oja, E.: Linear and nonlinear projective nonnegative matrix factorization. IEEE Trans. Neural Netw. **21**, 734–749 (2010). https://doi.org/10.1109/TNN.2010.2041361
17. Kaczkurkin, A.N., et al.: Evidence for dissociable linkage of dimensions of psychopathology to brain structure in youths. Am. J. Psychiatry **176**, 1000–1009 (2019). https://doi.org/10.1176/appi.ajp.2019.18070835
18. Harrison, S.J., et al.: Large-scale Probabilistic Functional Modes from resting state fMRI. Neuroimage **109**, 217–231 (2015). https://doi.org/10.1016/j.neuroimage.2015.01.013
19. Zhang, Y., Brady, M., Smith, S.: Segmentation of brain MR images through a hidden Markov random field model and the expectation-maximization algorithm. IEEE Trans. Med. Imaging **20**, 45–57 (2001). https://doi.org/10.1109/42.906424
20. Alfaro-Almagro, F., et al.: Image processing and Quality Control for the first 10,000 brain imaging datasets from UK Biobank. Neuroimage **166**, 400–424 (2018). https://doi.org/10.1016/j.neuroimage.2017.10.034
21. Boutsidis, C., Gallopoulos, E.: SVD based initialization: A head start for nonnegative matrix factorization. Pattern Recognit. **41**, 1350–1362 (2008). https://doi.org/10.1016/j.patcog.2007.09.010
22. Ha, S.M., Bani, A., Sotiras, A.: Scalable NMF via linearly optimized data compression. In: Medical Imaging 2023: Image Processing, pp. 170–176. SPIE (2023). https://doi.org/10.1117/12.2654282
23. Papenberg, M., Klau, G.W.: Using anticlustering to partition data sets into equivalent parts. Psychol. Methods **26**, 161–174 (2021). https://doi.org/10.1037/met0000301

24. Crouse, D.F.: On implementing 2D rectangular assignment algorithms. IEEE Trans. Aerosp. Electron. Syst. **52**, 1679–1696 (2016). https://doi.org/10.1109/TAES.2016.140952
25. Alexander-Bloch, A., Giedd, J.N., Bullmore, E.: Imaging structural co-variance between human brain regions. Nat. Rev. Neurosci. **14**, 322–336 (2013). https://doi.org/10.1038/nrn 3465
26. Pang, J.C., et al.: Geometric constraints on human brain function. Nature **618**, 566–574 (2023). https://doi.org/10.1038/s41586-023-06098-1
27. Bijsterbosch, J.D., et al.: The relationship between spatial configuration and functional connectivity of brain regions. eLife **7**, e32992 (2018). https://doi.org/10.7554/eLife.32992
28. Harrison, S.J., et al.: Modelling subject variability in the spatial and temporal characteristics of functional modes. Neuroimage **222**, 117226 (2020). https://doi.org/10.1016/j.neuroimage. 2020.117226
29. Sydnor, V.J., et al.: Neurodevelopment of the association cortices: Patterns, mechanisms, and implications for psychopathology. Neuron **109**, 2820–2846 (2021). https://doi.org/10.1016/j. neuron.2021.06.016

Author Index

A

Aktar, Mumu 46
Ambroise, Corentin 102

B

Bani, Abdalla 163
Bazzano, Lydia 67
Bentley, Paul 153
Benet Nirmala, Anoop 143
Bieniek, Kevin 143
Bijsterbosch, Janine 163
Birk, Florian 23, 123

C

Cai, Zhuotong 34
Carmichael, Owen T. 14, 67
Carvalho, Regiane 112
Charisis, Sokratis 143
Chen, Fuyao 34
Chuang, Kai-Cheng 67

D

Das, Sandhitsu R. 3
de Paiva, Joselisa 112
Duchesnay, Edouard 102
Dufumier, Benoit 102
Dugan, Reagan 14
Duncan, James S. 34, 79, 133
Duong, Michael Tran 3
Dvornek, Nicha C. 79, 133

E

Earnest, Tom 163

F

Fadaee, Elyas 143
Fang, Zhenghan 56
Filho, Fabiano 112
Fox, Peter 143
Frouin, Vincent 102

G

Galstyan, Aram 91
Grigis, Antoine 102
Gupta, Abha R. 133

H

Ha, Sung Min 163
Habes, Mohamad 143
Haddad, Elizabeth 91
Heczko, Samuel 123
Honnorat, Nicolas 143

J

Jahanshad, Neda 91

K

Kersten-Oertel, Marta 46
Khandelwal, Pulkit 3
Kirby, Krystal 67

L

Lenzini, Petra 163
Li, Jinqi 143
Li, Karl 143
Li, Xu 56
Lieffrig, Eléonore V. 34
Lohmann, Gabriele 23, 123
Loureiro, Rafael 112
Lu, Yihuan 34
Lyu, Xueying 3

M

Mahler, Lucas 23, 123
Marcus, Adam 153
Marmarelis, Myrl G. 91
Mendes, Giovanna 112

N

Nasrallah, Ilya M. 3
Nir, Talia M. 91

© The Editor(s) (if applicable) and The Author(s), under exclusive license
to Springer Nature Switzerland AG 2023
A. Abdulkadir et al. (Eds.): MLCN 2023, LNCS 14312, pp. 173–174, 2023.
https://doi.org/10.1007/978-3-031-44858-4

O
Olegário, Tayran   112
Onofrey, John A.   34

P
Paulo, Artur   112
Pinto, Bruna   112

R
Ramakrishnapillai, Sreekrishna   67
Rashid, Tanweer   143
Reis, Eduardo   112
Reis, Márcio   112
Ribeiro, Guilherme   112
Richardson, Timothy E.   143
Rittner, Letícia   112
Rivaz, Hassan   46
Rueckert, Daniel   153

S
Santos, Paulo   112
Scheffler, Klaus   23, 123
Seshadri, Sudha   143
Shin, Hyeong-Geol   56
Silva, Camila   112
Sotiras, Aristeidis   163
Staib, Lawrence H.   79
Steeg, Greg Ver   91
Steiglechner, Julius   23, 123

Sulam, Jeremias   56
Sullivan, Catherine   133

T
Toyonaga, Takuya   34

V
Van Gemmert, Arend W. A.   67
Vemula, Aishwarya   143

W
Walker, Jamie M.   143
Wang, Di   143
Wang, Jiyao   79
Wang, Qi   23, 123
Wolk, David A.   3

X
Xiao, Yiming   46
Xie, Long   3
Xin, Jingmin   34

Y
You, Chenyu   34
Yushkevich, Paul A.   3

Z
Zeng, Tianyi   34
Zhang, Jiazhen   34
Zheng, Nanning   34
Zijl, Peter van   56

Printed in the United States
by Baker & Taylor Publisher Services